This student workbook is intended to reinforce your learning of the content in the three units on the new single award GCSE Science specification (2101), from Edexcel.

Every worksheet is cross-referenced to the revision guide, *The Essentials of Edexcel GCSE Science*, published by Lonsdale.

The worksheets for each unit are divided into four topics to correspond with the specification and revision guide, and provide a clear, manageable structure to your revision. They focus on the material which is externally assessed (i.e. tested under exam conditions). They do not cover the practical skills assessment and assessment activities, which are marked by your teacher.

The questions and activities are designed to strengthen your understanding of the subject-specific content and how science works. They are varied in style to keep your study interesting, and there is limited space for each question so you will have to think carefully about your answers.

At the end of each topic there is an activity, e.g. a crossword or wordsearch, so that you can test your knowledge of the key words and concepts in that topic.

> HT Throughout this workbook, questions covering content that is limited to the Higher Tier exam papers appear inside a shaded box, clearly labelled with the symbol HT.

At the end of the book, you will find a detailed periodic table, which will provide a useful reference as you work through the chemistry worksheets.

## A Note to Teachers

The pages in this workbook can be used as...

- classwork sheets – students can use the revision guide to answer the questions
- harder classwork sheets – pupils study the topic and then answer the questions without using the revision guide
- easy-to-mark homework sheets – to test pupils' understanding and reinforce their learning
- the basis for learning homework tasks which are then tested in subsequent lessons
- test materials for topics or entire units
- a structured revision programme prior to the exams.

Answers to these worksheets are available to order.

**ISBN: 978-1-905129-64-5**

Published by Lonsdale, A division of Huveaux Plc

**Project Editor:** Rebecca Skinner

**Consultants:** Aleksander Jedrosz, John Watts, Susan Loxley

**Cover and Concept Design:** Sarah Duxbury

**Design:** Little Red Dog Design



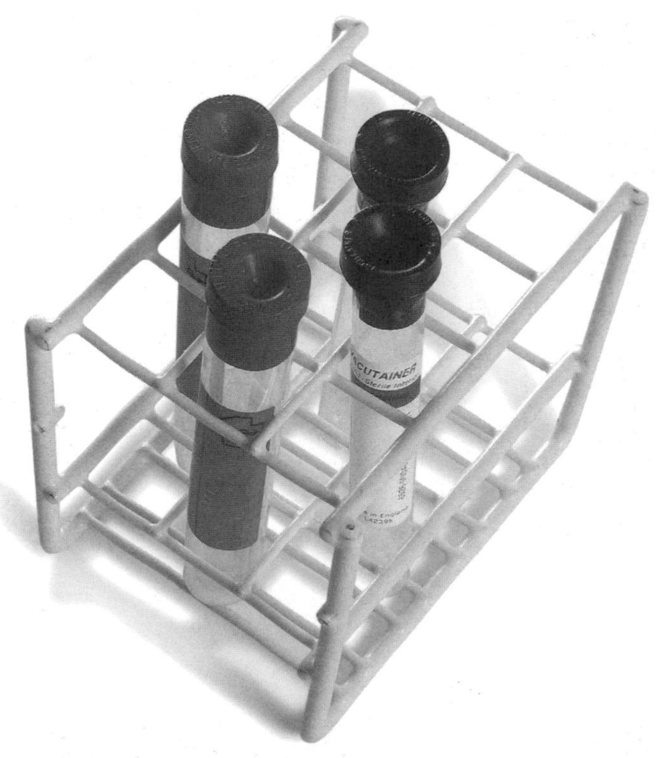

# Contents

# Contents

# Environment

1 a) Explain what a food chain shows.

..........

b) What is a pyramid of biomass?

..........

..........

c) Draw a pyramid of biomass for the following food chain:

Rose Bush ⟶ Greenfly ⟶ Blue Tit ⟶ Hawk

2 The diagram below shows how 1000J of energy flows through a food chain.

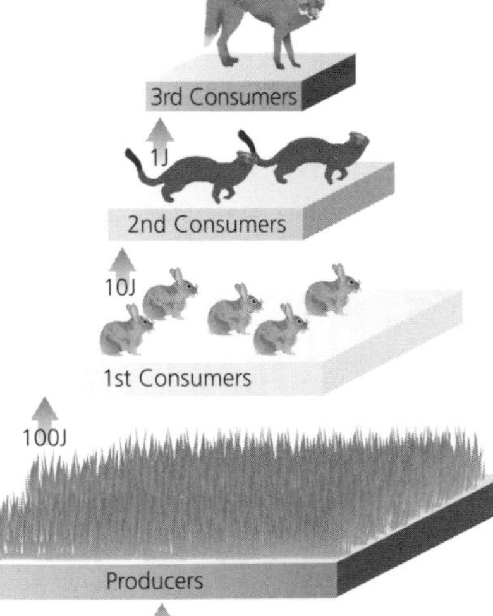


a) Name the process by which the grass captures energy from the Sun.

..........

b) What life process in the rabbits results in the transfer of heat to the surroundings?

..........

c) In this food chain, what percentage of the original energy from the Sun does the fox receive?

Energy received = .......... %

d) Describe the ways in which energy is lost at each stage in the food chain.

..........

..........

..........

1 **a)** In terms of energy, why is it more efficient for a human to be vegetarian?

**b)** Why do you think that most of the animals eaten by humans are herbivores, rather than carnivores?

2 At a supermarket, a 500g bag of rice costs 55p. Sirloin steak costs £7.49 for 500g. In terms of energy and biomass, explain the difference in price in as much detail as you can.

3 The size of a population of swallows, found in Devon, was recorded every other summer between 1990 and 1998.

| Year | No. of Swallows |
|---|---|
| 1990 | 956 |
| 1992 | 918 |
| 1994 | 876 |
| 1996 | 834 |
| 1998 | 797 |

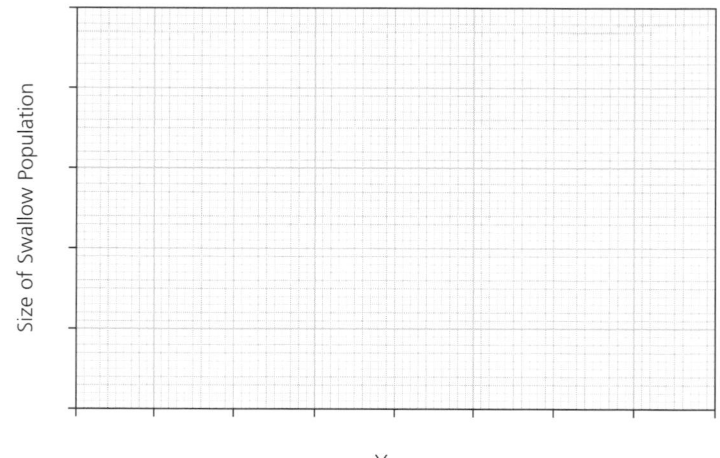


**a)** Draw a graph to show the data in the table.

**b)** Estimate what the size of the swallow population was in the year 2000.

**c)** Write down one possible reason to explain why the population of swallows was decreasing over this period.

# Environment

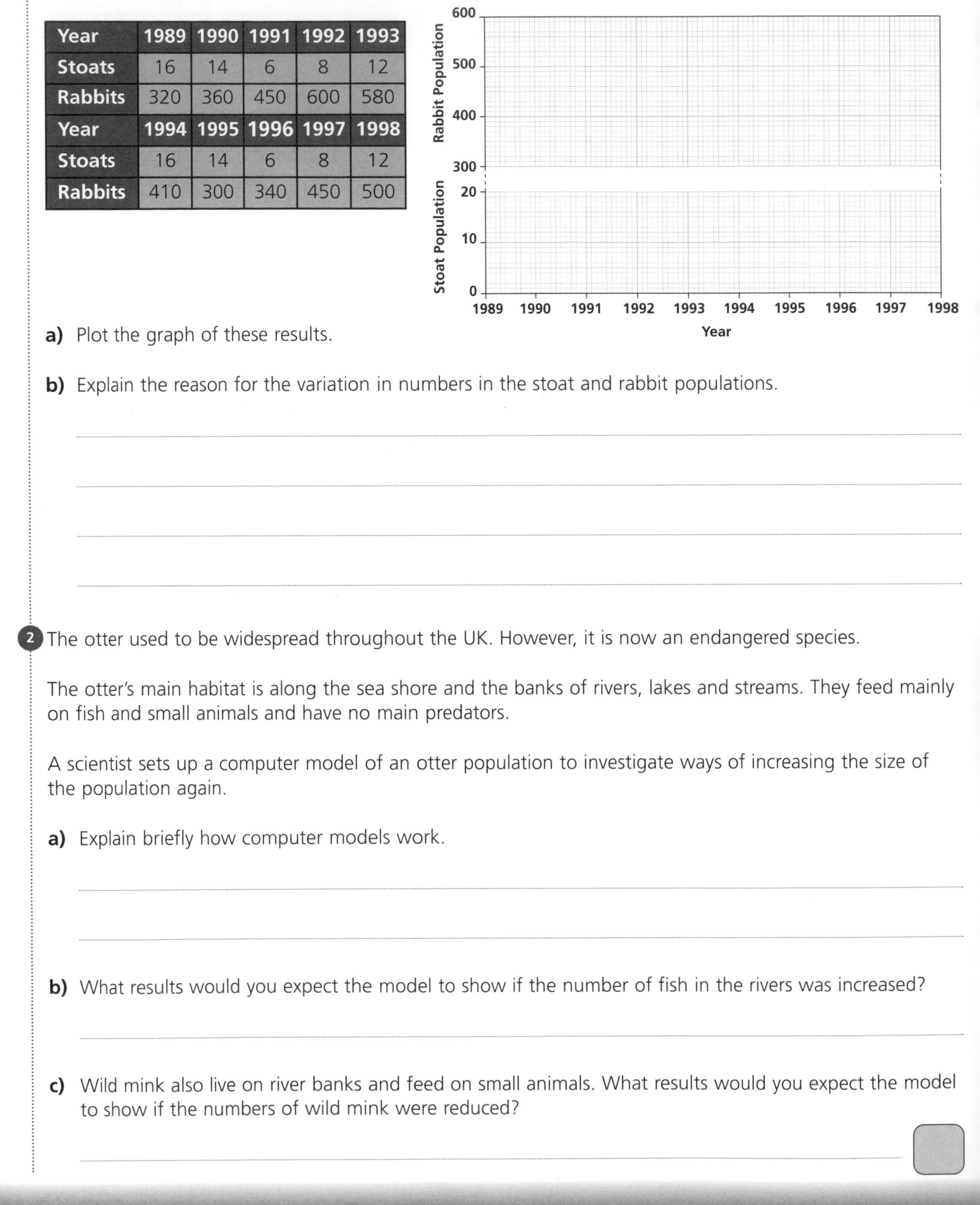

1 The following information on the population of stoats and rabbits in a particular area was obtained over a period of 10 years.

| Year | 1989 | 1990 | 1991 | 1992 | 1993 |
|---|---|---|---|---|---|
| Stoats | 16 | 14 | 6 | 8 | 12 |
| Rabbits | 320 | 360 | 450 | 600 | 580 |
| **Year** | **1994** | **1995** | **1996** | **1997** | **1998** |
| Stoats | 16 | 14 | 6 | 8 | 12 |
| Rabbits | 410 | 300 | 340 | 450 | 500 |

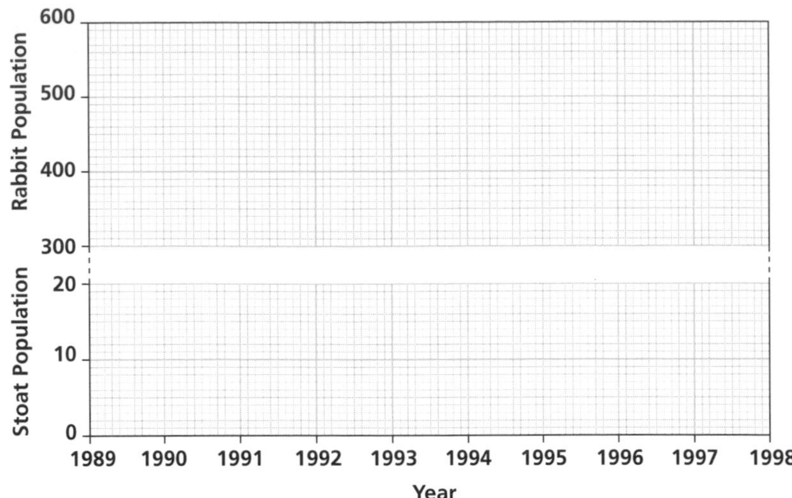


**a)** Plot the graph of these results.

**b)** Explain the reason for the variation in numbers in the stoat and rabbit populations.

2 The otter used to be widespread throughout the UK. However, it is now an endangered species.

The otter's main habitat is along the sea shore and the banks of rivers, lakes and streams. They feed mainly on fish and small animals and have no main predators.

A scientist sets up a computer model of an otter population to investigate ways of increasing the size of the population again.

**a)** Explain briefly how computer models work.

**b)** What results would you expect the model to show if the number of fish in the rivers was increased?

**c)** Wild mink also live on river banks and feed on small animals. What results would you expect the model to show if the numbers of wild mink were reduced?

1 Complete the table below to compare the different processes that can change the characteristics of a species over time.

| Method | Description | How it Happens |
|---|---|---|
| a) | Individuals that are better adapted to the environment survive and reproduce. | |
| b) | | Human Intervention |
| c) Genetic Engineering | Genes from one organism are transferred into another to introduce new characteristics. | |

2 Which process do you think was responsible for each of the following changes?

**a)** A new variety of rice has been developed that contains beta-carotene (found in carrots), which is needed by the body to produce Vitamin A.

**b)** Mexican Cave Fish live in a very dark environment. Over many generations their eyes have disappeared.

**c)** The wheat grown on modern farms has a far higher yield of grain per stalk than the wheat grown by our ancestors three hundred years ago.

3 **a)** State two advantages of selective breeding.

**i)**

**ii)**

**b)** Name two features you would select if you were breeding racehorses.

**i)**

**ii)**

**c)** Use your imagination to create a completely new variety of organism that could possibly be produced by selective breeding or genetic engineering, and would be useful in modern farming. Try to make it as believable as possible.

# Environment

**1 a)** Look up 'organic farming' in a dictionary. Summarise what it means in your own words.

**b)** List three key principles of organic farming.

**i)**

**ii)**

**iii)**

**c)** In a supermarket, royal gala apples cost £1.18 per kg. Organic royal gala apples cost £4.34 per kg.

**i)** If you wanted to buy 500g of apples, how much more would it cost you to buy the organic variety?

**ii)** Explain, in as much detail as possible, why there is a difference in price.

**2** Humans may share a common ancestor with apes. Do you think this is possible? Explain your answer.

HT

**3** What four important observations led Charles Darwin to develop his theory of evolution?

**a)**

**b)**

**c)**

**d)**

**1** **a)** What are fossils?

**b)** Explain why the fossil record is incomplete.

**2** The following data shows the number of Peppered Moths spotted in a survey in the centre of Manchester before and after the Industrial Revolution.

| Month | Pre-Industrial Revolution | | Post-Industrial Revolution | |
|---|---|---|---|---|
| | Pale | Dark | Pale | Dark |
| **June** | 1261 | 102 | 87 | 1035 |
| **July** | 1247 | 126 | 108 | 1336 |
| **August** | 1272 | 93 | 72 | 1019 |

**a)** Calculate the mean number of each type of moth for each three-month period.

**i)** Pre-Industrial Revolution: Pale = .................... Dark = ....................

**ii)** Post-Industrial Revolution: Pale = .................... Dark = ....................

**b)** Explain why there were more pale moths than dark moths prior to the Industrial Revolution.

**c)** Explain why the number of dark moths increased significantly after the Industrial Revolution.

**d)** The Clean Air Act was introduced in the 1950s. How do you think this affected the environment in Manchester?

**e)** How would you expect this to affect the relative proportion of pale moths to dark moths?

# Environment

1 Name the five kingdoms used to classify all living organisms.

**a)** ..................

**b)** ..................

**c)** ..................

**d)** ..................

**e)** ..................

2 **a)** Why would grouping animals by colour not be a reasonable method of classification?

**b)** Why would grouping animals by name not be a reasonable method of classification?

3 If you found a new animal specimen in a tropical rainforest, what immediate questions would you ask to begin classifying it?

4 When classifying an organism, the groups gradually get smaller and smaller. As this happens, what would you expect to see in the organisms within the group?

5 How would a scientist prove that an organism he had discovered was a new species?

6 Name one organism that might be difficult to classify accurately, and explain your answer.

# Environment

1 Complete the crossword below.

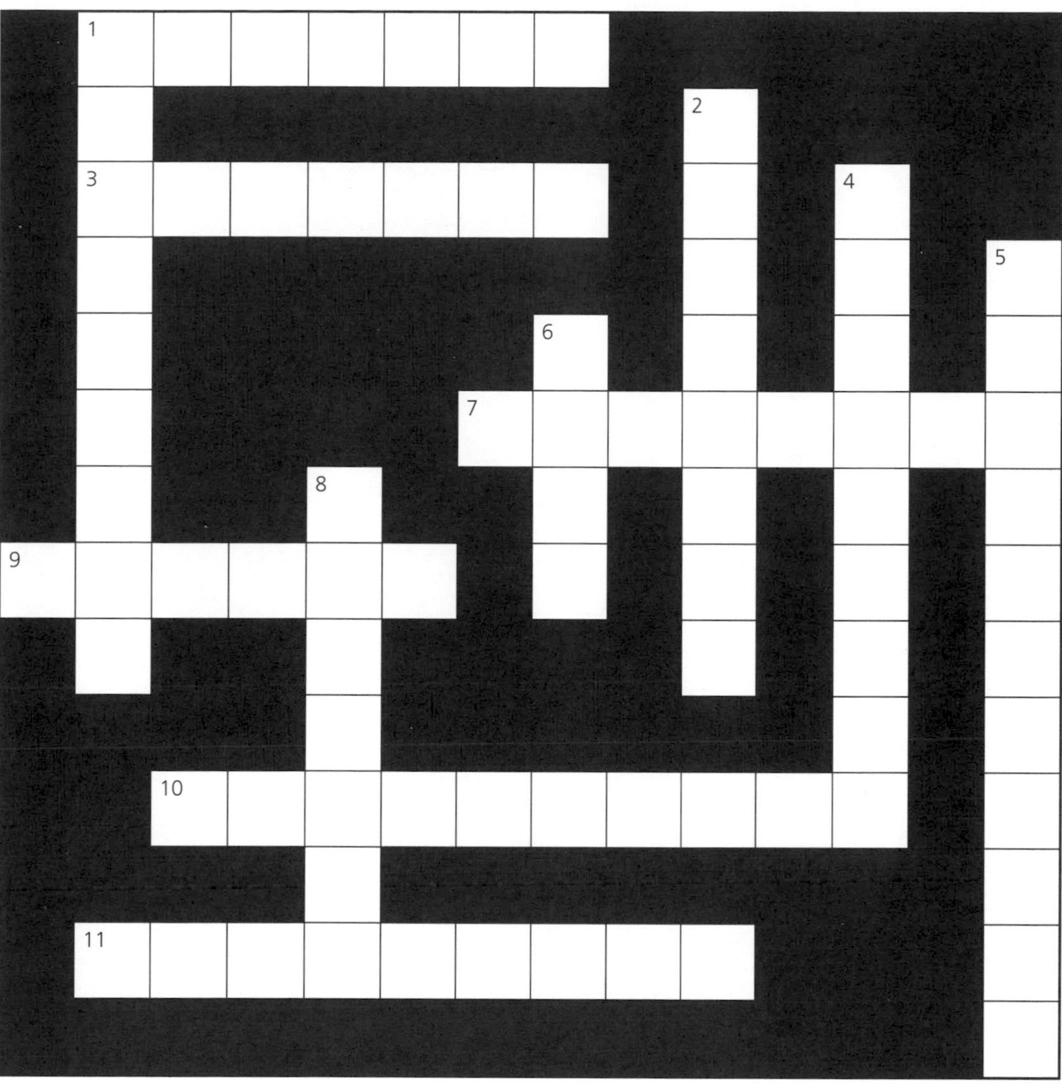


**Across**

1. Cease to exist (7)
3. Natural; from living matter (7)
7. A living thing (8)
9. The preserved remains of an organism (6)
10. A change or adjustment which improves an organism's ability to survive (10)
11. Describes all the organisms within an environment, and their relationships with each other and the environment (9)

**Down**

1. The process by which a species changes over many generations, to become better adapted for survival (9)
2. Describes a carnivorous animal, which hunts other animals for food (8)
4. Differences between individuals of the same species (9)
5. This occurs when organisms require the same resources to survive (11)
6. Describes an animal that is hunted by other animals for food (4)
8. The mass of organic material in an organism or population (7)

# Genes

1 a) Explain, as fully as you can, what genes are.

b) What are chromosomes?

c) How many chromosomes would you find in a human skin cell?

2 A DNA molecule consists of two strands linked together. Describe, with the help of a diagram, how they are linked.

3 Explain, in your own words, what genes do.

4 What was the purpose of the Human Genome Project?

HT 5 Some transgenic animals are produced as disease models; they exhibit disease symptoms so that potential treatments can be tested.

Explain what a transgenic animal is.

# Genes

1. What is gene therapy?

2. What is cystic fibrosis?

3. Describe, in general terms, how modified viruses are used to alleviate the symptoms of diseases such as cystic fibrosis.

4. **a)** During the treatment of cystic fibrosis by gene therapy, a number of steps are involved. Number the following stages **1** to **6**, to show the correct sequence.

   **i)** The viruses deliver new genetic material into the cells. ☐

   **ii)** The cells in the respiratory system reduce their production of mucus. ☐

   **iii)** New genetic material needed by the cystic fibrosis sufferer is inserted. ☐

   **iv)** The viral vector is inhaled. ☐

   **v)** Disease-causing genetic material is removed from viruses. ☐

   **vi)** Treatment has to be repeated to alter the genetic make-up of new cells. ☐

   **b)** Explain why the treatment outlined above does not stop the next generation (offspring) becoming victims of the same disorder.

# Genes

1 *Sexual reproduction promotes variation.* Explain this statement.

2 In terms of inheriting genes, how is asexual reproduction different from sexual reproduction?

3 Spider plants reproduce asexually by producing runners. Use the Internet, a text book, or another secondary source, to find two more examples of plants that reproduce in this way.

**a)** ............ **b)** ............

4 There are several methods of cloning. Dolly the sheep was cloned from an adult cell, taken from the udder of a ewe. Name one other method.

5 **a)** Suggest one way in which cloning techniques could be used to benefit society.

**b)** Suggest two disadvantages or concerns about cloning, especially animals.

**i)** ............

**ii)** ............

HT **c)** Read the headline and strap line from a newspaper article below.

**'Designer Baby' Ban Quashed**

A ban on a couple creating a baby to help save their child has been overturned by the Court of Appeal.

What is meant by the term 'designer baby'?

# Genes

**1 a)** What is variation?

**b)** Name the two factors that can cause variation, and provide an example for each one.

**i)** Factor: ........ Example: ........

**ii)** Factor: ........ Example: ........

**c)** Below is a list of features George uses to describe himself. For each one, state whether it is **genetic, environmental** or a **combination** of both.

**i)** 2.01m tall ........ **ii)** blue eyes ........

**iii)** blonde hair ........ **iv)** scar on his forehead ........

**v)** speaks German ........ **vi)** a bit overweight ........

HT

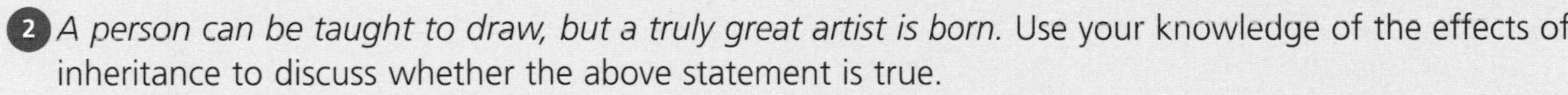

**2** *A person can be taught to draw, but a truly great artist is born.* Use your knowledge of the effects of inheritance to discuss whether the above statement is true.

**3** The graphs below show the birth length of babies from two different groups of mothers.

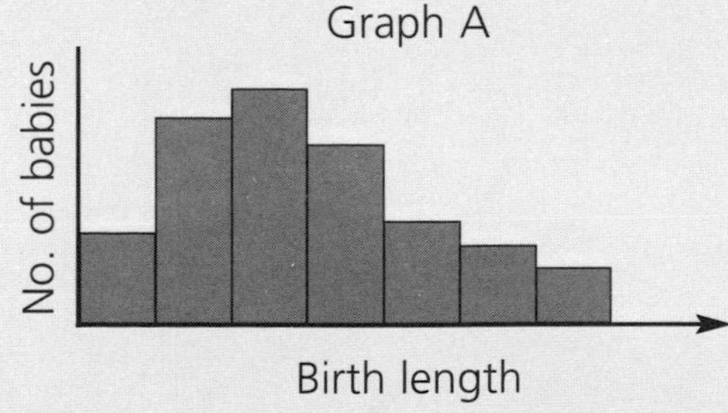


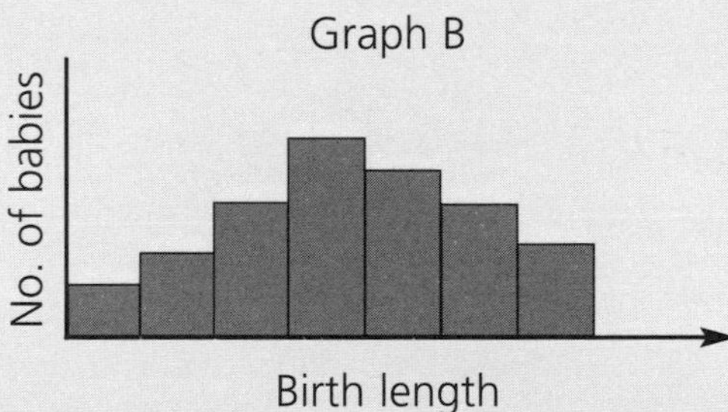


Graph A shows the data from mothers who are smokers. How has their mothers' smoking affected the birth weight of this group of babies?

# Genes

1 Genes can have alternative forms, called alleles. Fill in the missing words to complete the following sentences about alleles.

**a)** An allele that controls the development of a characteristic whenever it is present is called a ........................ allele.

**b)** An allele that controls the development of a characteristic only when it is present on both chromosomes in a pair is called a ........................ allele.

2 What type of allele causes...

**a)** cystic fibrosis? ........................ **b)** Huntington's disease? ........................

3 If H is the dominant allele that causes Huntington's disease and h is the recessive allele ('healthy' allele), do the following individuals have Huntington's disease?

**a)** HH ........................ **b)** hh ........................ **c)** Hh ........................

4 **a)** If c is the recessive allele that causes cystic fibrosis and C is the dominant allele ('healthy' allele), do the following individuals have cystic fibrosis?

**i)** CC ........................ **ii)** cc ........................ **iii)** Cc ........................

**b)** What word is used to describe individuals who do not have cystic fibrosis, but do have one 'faulty' recessive allele? ........................

HT

5 State whether the following genotypes are **homozygous dominant, heterozygous** or **homozygous recessive.**

**a)** Tt ........................ **b)** BB ........................

**c)** ee ........................ **d)** EE ........................

**e)** Bb ........................ **f)** tT ........................

6 The gene that controls ear lobe type has two alleles. The allele for attached lobes is recessive (e) and the allele for unattached lobes is dominant (E).

Write down all the possible combinations and, for each one, state the outcome.

........................

........................

........................

HT 1 Complete these two different crosses between a brown-eyed and a blue-eyed parent (brown eyes are dominant).

**Brown eyes X Blue eyes**

**Parents** BB X bb

**Gametes**

**Offspring**

**Brown eyes X Blue eyes**

**Parents** Bb X bb

**Gametes**

**Offspring**

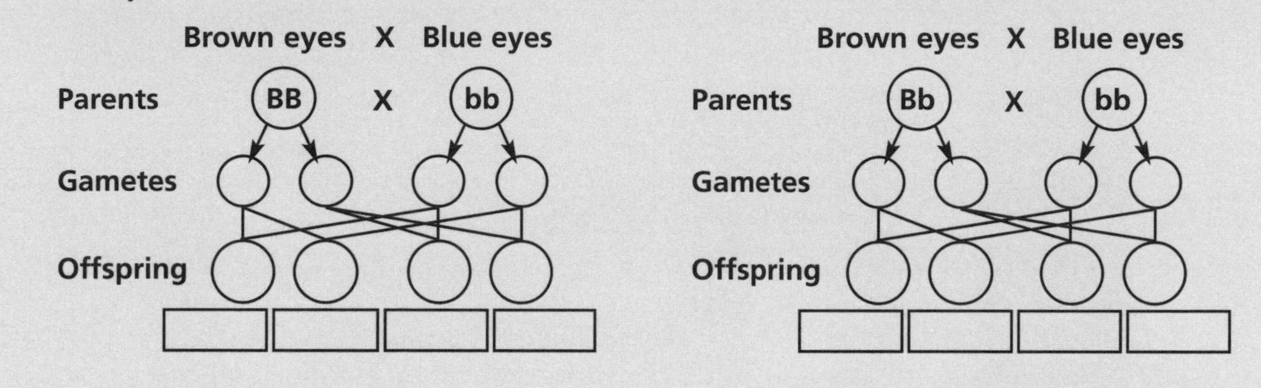

2 Explain how two parents with brown eyes could produce a child with blue eyes. Use a diagram to help you.

3 **a)** Complete the following diagrams to show how attached and unattached ear lobes might be inherited (unattached lobes are dominant).

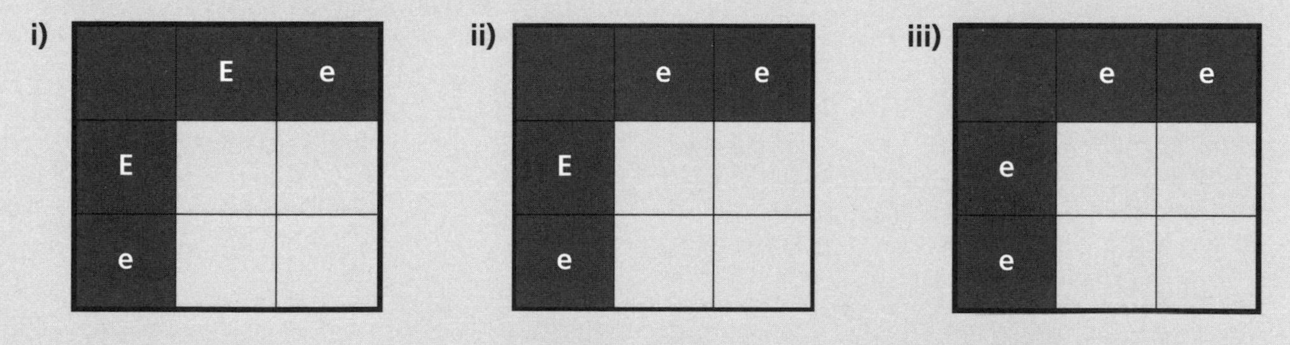

**i)**

| | E | e |
|---|---|---|
| E | | |
| e | | |

**ii)**

| | e | e |
|---|---|---|
| E | | |
| e | | |

**iii)**

| | e | e |
|---|---|---|
| e | | |
| e | | |

**b)** What are the percentage chances of producing a child with attached ear lobes from each of the crosses above?

**i)**

**ii)**

**iii)**

# Genes

**1** Complete the crossword below.

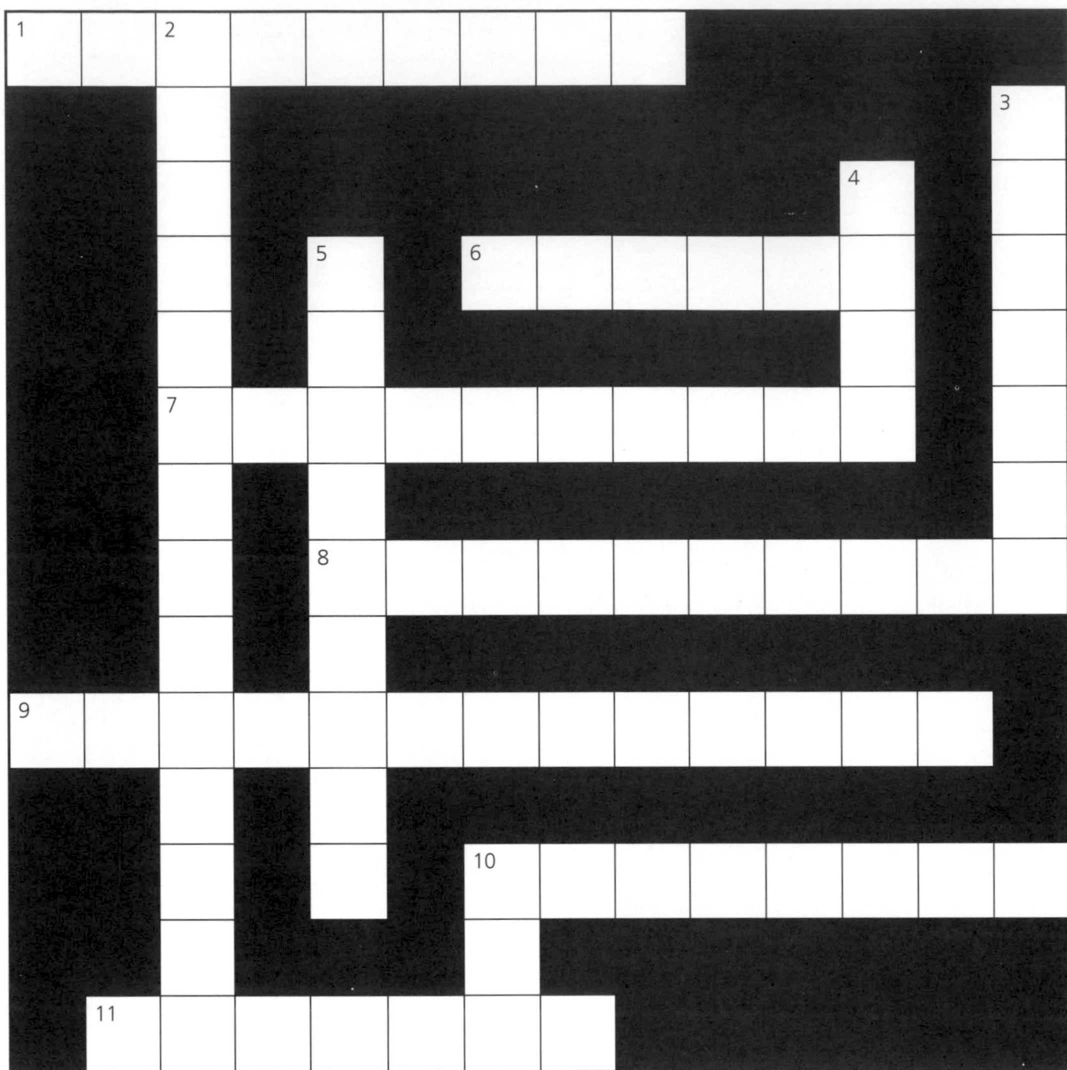


**Across**

1. Describes an allele that only controls a characteristic if present on both chromosomes in a pair (9)
6. An alternative form of a gene (6)
7. A string of genes; made from DNA (10)
8. Some white blood cells produce these when they detect antigens (10)
9. The stage at which gametes fuse together during sexual reproduction (13)
10. Describes an allele that controls a characteristic, even if present on only one chromosome in a pair (8)
11. Reproduction involving only one parent (7)

**Down**

2. A genetic disorder, which affects the cell membranes (6,8)
3. Sex cells (7)
4. A section of DNA, which codes for the synthesis of a protein (4)
5. The differences between individuals of the same species (9)
10. The chemical from which chromosomes and genes are made (3)

# Electrical and Chemical Signals

1 Why is it an advantage for an organism to possess a nervous system?

2 Label the three parts of the central nervous system shown on the diagram below.

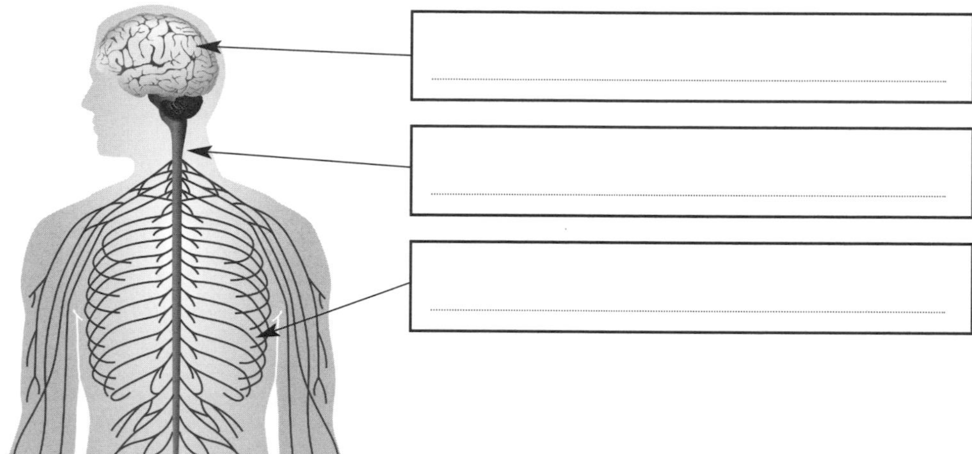

3 Complete the table below, naming the stimulus that the receptors in each organ respond to. The first one has been done for you.

| Organ | Stimulus |
|---|---|
| **a)** ears | Sound and balance |
| **b)** eyes | |
| **c)** skin | |
| **d)** tongue | |
| **e)** nose | |

4 Use a line to link each different part of the brain to the correct function.

| | |
|---|---|
| **a)** Cerebellum | Controls automatic actions, e.g. heart beat and breathing. |
| **b)** Medulla | Coordinates movement and balance. |
| **c)** Cerebral Hemisphere | Thought, memory and emotions. |

# Electrical and Chemical Signals

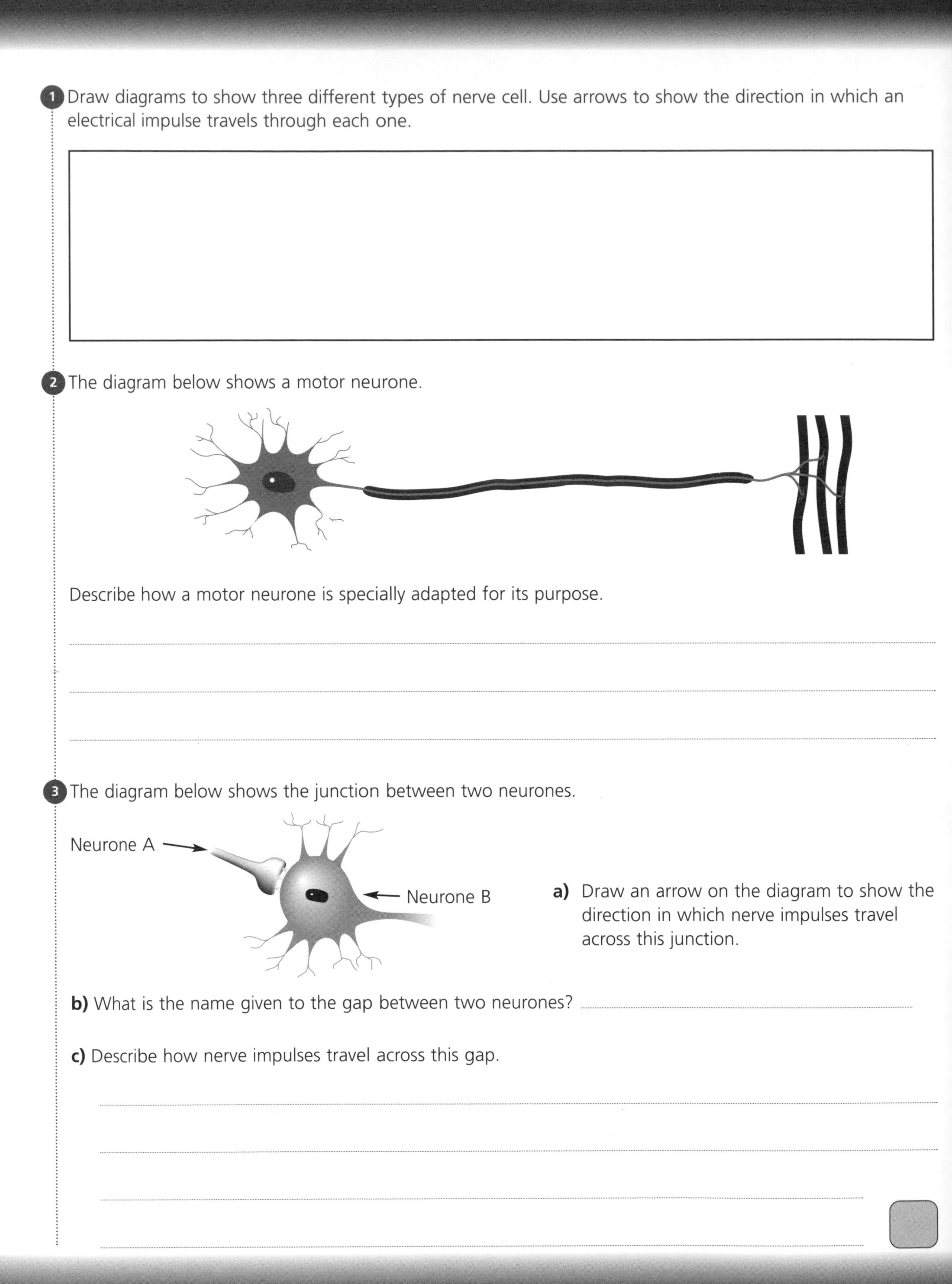

1 Draw diagrams to show three different types of nerve cell. Use arrows to show the direction in which an electrical impulse travels through each one.

2 The diagram below shows a motor neurone.

Describe how a motor neurone is specially adapted for its purpose.

3 The diagram below shows the junction between two neurones.

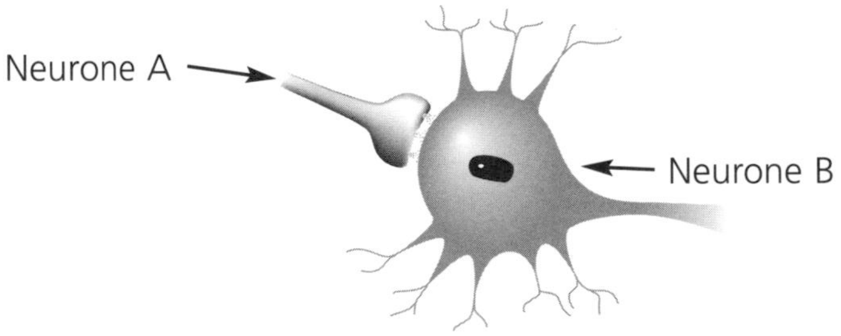

a) Draw an arrow on the diagram to show the direction in which nerve impulses travel across this junction.

b) What is the name given to the gap between two neurones?

c) Describe how nerve impulses travel across this gap.

# Electrical and Chemical Signals

1. Number the following parts of the nervous system **1** to **7** to show the pathway that a nervous impulse follows through the body in response to a stimulus.

   **a)** Effector (e.g. muscle) ☐

   **b)** Synapse between sensory neurone and relay neurone ☐

   **c)** Receptor (in sense organ) ☐

   **d)** Motor neurone ☐

   **e)** Relay neurone ☐

   **f)** Synapse between relay neurone and motor neurone ☐

   **g)** Sensory neurone ☐

2. In the kitchen, Paul reaches out his hand to take a saucepan off the hob. 'Ow, it's hot!' he says, pulling his hand away.

   **a)** Paul's reaction was a reflex response. What does this term mean?

   ..........................................................................................

   ..........................................................................................

   **b)** The diagram below shows the reflex arc involved in this action. Name the parts labelled A, B, C and D.

   A = ..........................................

   B = ..........................................

   C = ..........................................

   D = ..........................................

   A
   B
   D
   C
   Spinal Cord

   **c)** Describe the nervous pathways of this reflex arc.

   ..........................................................................................

   ..........................................................................................

   ..........................................................................................

   ..........................................................................................

# Electrical and Chemical Signals

1 a) Describe how their eyes respond to bright light when a person walks out of a darkened cinema.

b) Explain your answer to part **a)** in terms of the muscles in the iris.

c) Draw diagrams in the space below to show the iris and pupil of an eye in **i)** dim light, **ii)** moderate light, and **iii)** bright light. Label them accordingly.

d) Describe, in your own words, the nervous pathway that controls this response.

2 a) The human eye can adjust to focus on objects at a wide range of different distances. What is this ability called?

b) Explain, in as much detail as you can, how the eye focuses on a near object. Use a diagram to help you.

# Electrical and Chemical Signals

1 Name the four components of blood.

a) ........................ b) ........................

c) ........................ d) ........................

2 Blood carries oxygen around the body.

a) Which part of the blood carries the oxygen? ........................

b) What features make it well adapted to this function?

........................

........................

c) Describe how it carries the oxygen from the lungs to the organs in the body.

........................

........................

3 In addition to its nervous system, the human body uses chemicals to help control its functions.

a) What are these chemicals called? ........................

b) Where are these chemicals produced? ........................

c) How do these chemicals travel around the body? ........................

d) Name three different chemical messengers and, for each one, state where it is produced.

i) ........................

ii) ........................

iii) ........................

HT 4 Oestrogen and progesterone help to control a woman's menstrual cycle. Describe the changes that occur to the levels of oestrogen during the menstrual cycle. Link the changes to the thickness of the uterus lining.

........................

........................

# Electrical and Chemical Signals

1 Synthetic hormones can be used to control a woman's fertility artificially.

**a)** What does 'synthetic' mean?

**b)** What type of synthetic hormones are used in…

**i)** The contraceptive pill?

**ii)** Female fertility drugs?

2 **a)** Describe one situation where a couple might try IVF treatment to improve their chance of conceiving.

**b)** Suggest one reason why a couple might prefer to try IVF treatment rather than adopt a baby.

**c)** Suggest one reason why a couple might prefer to adopt a baby rather than try IVF treatment.

3 Below is some additional information about IVF treatment.

In normal pregnancies, no more than 2% of women have a multiple birth, i.e. give birth to twins, triplets etc. With IVF treatment, the chance of having a multiple birth rises to 25%.

**a)** Describe one problem that a multiple birth might cause for a couple.

**b)** If IVF treatment becomes more common, the number of multiple births will increase. What effect would this have on the population?

**c)** Describe one way in which this could impact on society.

# Electrical and Chemical Signals

1 **a)** A malfunction of which organ in the body causes diabetes? ..........

**b)** How does this organ cause diabetes?

..........

**c)** What effect does this have on the person's blood?

..........

2 The graph shows the blood sugar level for a person with diabetes who didn't take insulin for a 12 hour period.

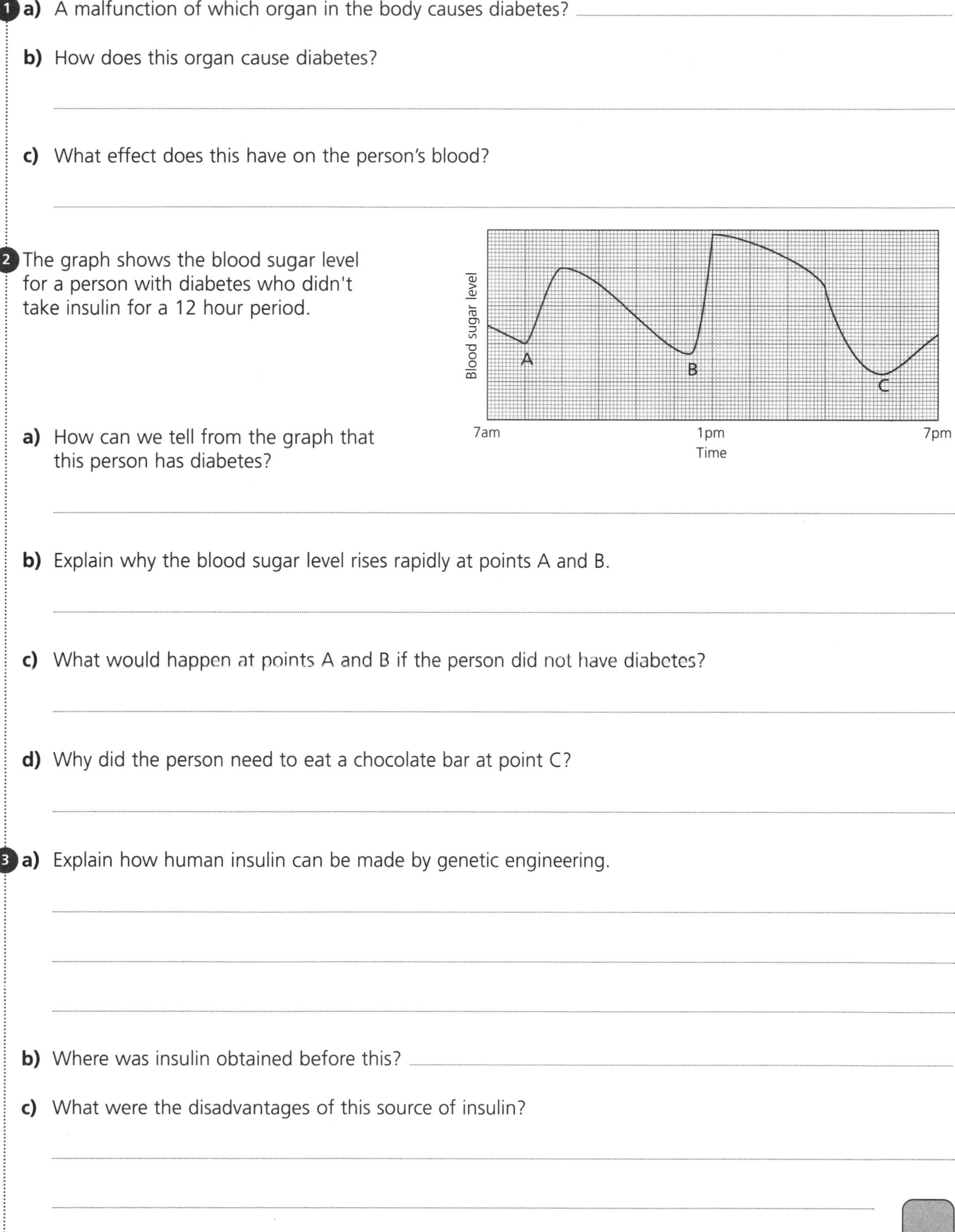


**a)** How can we tell from the graph that this person has diabetes?

..........

**b)** Explain why the blood sugar level rises rapidly at points A and B.

..........

**c)** What would happen at points A and B if the person did not have diabetes?

..........

**d)** Why did the person need to eat a chocolate bar at point C?

..........

3 **a)** Explain how human insulin can be made by genetic engineering.

..........

..........

..........

**b)** Where was insulin obtained before this? ..........

**c)** What were the disadvantages of this source of insulin?

..........

..........

# Electrical and Chemical Signals

1 Complete the crossword below.

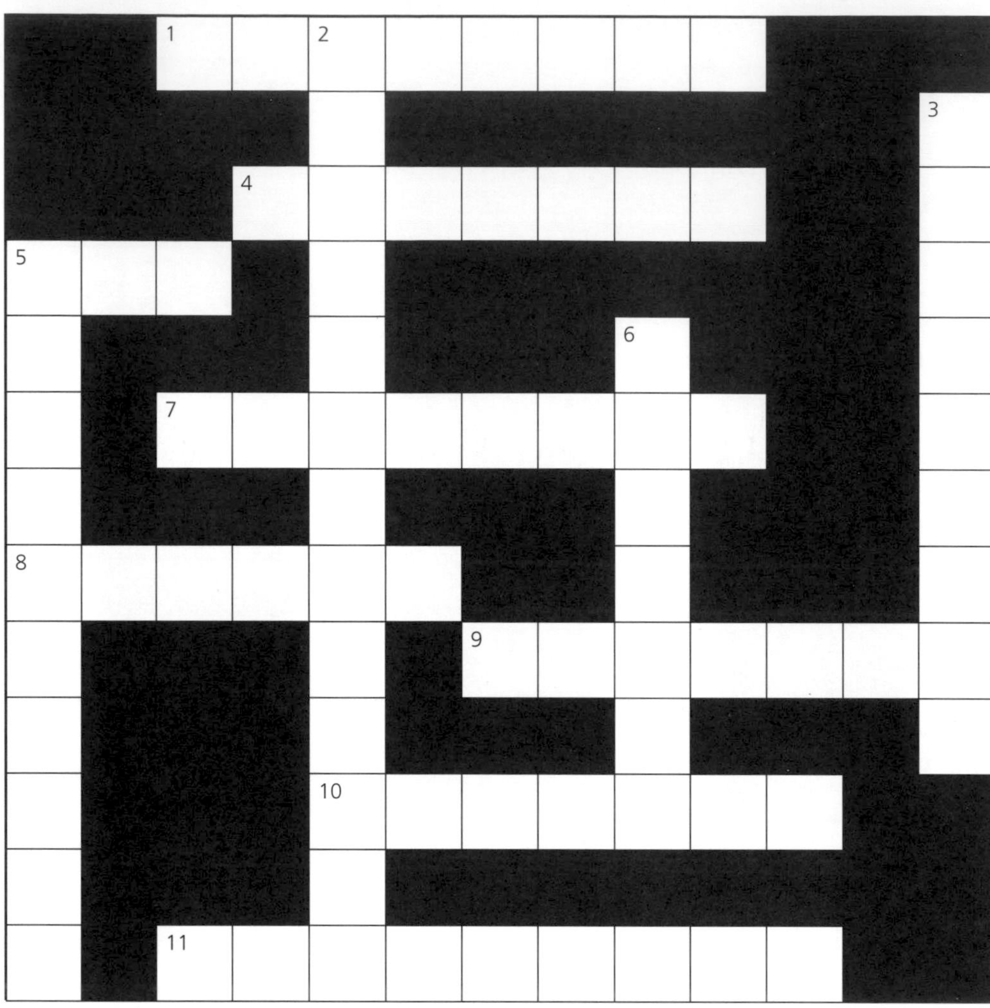


**Across**

1. This senses a stimulus and triggers a nervous response (8)
4. An example of 9 across; helps regulate blood sugar concentration (7)
5. A fertility treatment, in which fertilisation takes place outside the mother's body (3)
7. A disease caused when the pancreas fails to produce enough insulin (8)
8. An automatic response or action (6)
9. A chemical messenger, produced by an endocrine gland (7)
10. An electrical signal, transmitted by nerve cells (7)
11. A hormonally controlled cycle that takes place in a woman's reproductive system (9)

**Down**

2. Birth control (13)
3. The hormone that causes the lining of the uterus to start to thicken during 11 across (9)
5. An involuntary action which causes the pupil of the eye to open and close in response to light (4, 6)
6. The part of the nervous system which coordinates nervous responses (7)

# Use, Misuse and Abuse

1 a) Which substance in tobacco is addictive?

b) Which substance in tobacco can cause cancer?

c) Name three other diseases caused by smoking.

i) ii) iii)

d) What are the harmful effects of smoking on the respiratory system?

e) What are the harmful effects of smoking on the circulatory system?

2 a) What effect does moderate alcohol consumption have on the body?

b) Give three reasons why driving under the influence of alcohol is inadvisable.

i)

ii)

iii)

c) State two long-term health problems associated with excess alcohol consumption.

i) ii)

3 a) What are solvents?

b) Name three problems caused by solvent abuse.

i) ii) iii)

# Use, Misuse and Abuse

1 a) Describe how caffeine affects the nervous system.

b) What effect does this have on behaviour?

2 In terms of the body's nervous system, explain how paracetamol stops you from feeling pain.

3 Barbiturates are useful sedatives. However, in recent years, doctors have started to use them much less. Why do you think this is?

4 In addition to the effects of the drug on the physical and mental health of addicts, there are other risks associated with heroin use. Suggest one.

5 List three things that cannabinoids and opiates have in common.

a)

b)

c)

6 The wordsearch alongside contains 7 words, which describe some of the possible effects that drugs can have on your body and behaviour. Draw a line through each of the 7 words.

| | | | | | | | | |
|---|---|---|---|---|---|---|---|---|
| E | X | H | A | U | S | T | E | D |
| I | A | V | Y | S | W | O | R | D |
| L | R | E | D | O | R | E | K | E |
| U | M | R | Y | S | J | I | U | S |
| P | X | S | I | T | C | H | Y | U |
| S | T | A | Z | T | C | F | Q | F |
| E | U | Y | X | T | A | O | U | N |
| T | M | P | I | P | H | B | G | O |
| B | Z | W | C | A | I | H | L | C |
| A | G | G | R | E | S | I | V | E |

# Use, Misuse and Abuse

1 a) What is a pathogen?

b) Name the three main types of pathogen.

i) ii) iii)

2 Pathogens can be transmitted through direct contact and indirect contact.

a) Which of these methods of transmission is the most common?

b) Name three different types of direct contact, and include a brief explanation of each one.

i)

ii)

iii)

c) What is a vector?

3 The body uses physical and chemical barriers to stop microorganisms from getting in. Unscramble the letters below to find five different types of barrier. For each one, briefly explain how it works.

a) DOLOB CLOGTINT

b) LANAS RASHI

c) MYSLYZOE

d) INKS

e) ILIAC

# Use, Misuse and Abuse

1 After two weeks in her first term of teaching, Miss Barton suffered a cold and a sore throat.

**a)** Explain what caused Miss Barton to feel ill.

**b)** All the other teachers said that when they first started teaching they seemed to get ill more often, but after their first year the frequency decreased. Explain why this happened.

**c)** Miss Barton recovered from her cold without taking any medicines. Explain how.

2 **a)** Name the two types of white blood cell.

**i)** ..................... **ii)** .....................

**b)** Explain the difference between these two types of white blood cell.

3 **a)** There are some diseases that people only ever get once. In terms of the white blood cells, explain why they do not get the same disease a second time.

**b)** What is this called? .....................

# Use, Misuse and Abuse

1 **a)** Use the Internet, a text book, or another secondary source, to find out some of the symptoms of tuberculosis. List four below.

**i)** ..................... **ii)** .....................

**iii)** ..................... **iv)** .....................

**b)** What type of pathogen causes TB? .....................

**c)** What part of the body does it usually affect? .....................

2 **a)** TB is highly contagious. Describe how it is transmitted from one person to another.

**b)** In the 19th century, the average Briton could only expect to live into their 50s, and many didn't even make it to adulthood. TB was responsible for approximately 25% of all deaths in Europe at that time.

The number of TB-related deaths fell dramatically throughout the 20th century, even before the vaccine was first used in 1953. Suggest one possible reason for this.

3 **a)** Why do you think TB is still common in developing countries?

**b)** Over the past ten years, there has been a rise in the number of TB cases in the UK again. Suggest one possible reason for this.

4 New drugs have to be tested before they can be approved for use. Below are the three main stages of testing. Use a line to link each stage to the correct description.

| | |
| --- | --- |
| **a)** Laboratory tests | Test the drug's effectiveness, and compare it with existing treatments. |
| **b)** Trials involving healthy volunteers | Test whether the drug is toxic. |
| **c)** Trials involving patients with the disease | Test the drugs for possible side effects. |

1 Complete the crossword below.

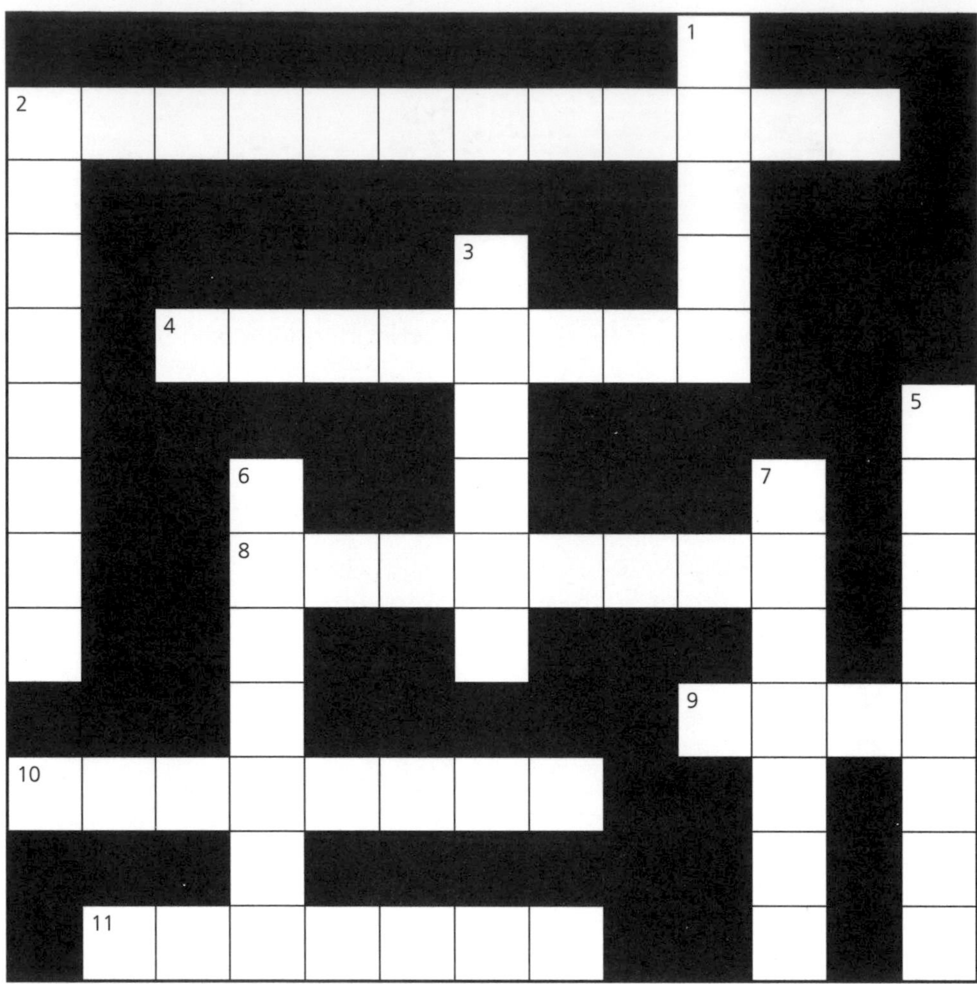


**Across**

2. Red, hot and painful swelling (12)
4. A single-celled organism with no nucleus (8)
8. A chemical found in tears, which destroys bacteria (8)
9. A chemical that alters the way the body works (4)
10. A disease-causing microorganism (8)
11. A substance that can dissolve other substances (7)

**Down**

1. Tiny, hair-like structures found on the surface of cells that line the trachea (5)
2. A natural or acquired resistance to a specific disease (8)
3. Something that acts as a vehicle or carrier (6)
5. When these are detected, white blood cells produce antibodies (7)
6. A common, legal drug, which acts as a sedative (7)
7. A nerve cell (7)

# Patterns in Properties

1 a) Complete the table about atomic particles below.

| Atomic Particle | Relative Mass | Relative Charge |
|---|---|---|
| | | +1 |
| | 1 | |
| | 0 (nearly) | |

b) Describe the structure of an atom in terms of these particles.

..........

..........

2 a) The letters A, B, C, D, E and F below represent six different elements. For each one write down **i)** the atomic number, **ii)** the mass number, **iii)** the number of protons in one atom, **iv)** the number of neutrons in one atom. (A, B, C, D, E and F are not their chemical symbols).

| | $^{12}_{6}A$ | $^{9}_{4}B$ | $^{19}_{9}C$ | $^{11}_{5}D$ | $^{28}_{14}E$ | $^{40}_{18}F$ |
|---|---|---|---|---|---|---|
| **i)** Atomic Number | | | | | | |
| **ii)** Mass Number | | | | | | |
| **iii)** No. of Protons | | | | | | |
| **iv)** No. of Neutrons | | | | | | |

b) Use the periodic table at the back of this book to identify the elements A ,B, C, D, E, and F.

A= .......... B= .......... C= .......... D= .......... E= .......... F= ..........

c) Explain what is meant by....

**i)** Atomic number ..........

**ii)** Mass number ..........

3 a) What is a molecule? ..........

b) Give one example of a diatomic molecule and draw a simple diagram to illustrate it.

..........

..........

# Patterns in Properties

1 In the following word equation, write **reactant** or **product** below each substance to show what it is.

Iron + Sulphur ⟶ Iron sulphide

.................. .................. ..................

2 Put a ring around the correct word or phrase in each of these sentences about chemical reactions.

**a)** During an exothermic reaction heat energy is **released / taken in**.

**b)** During an exothermic reaction there is a **rise / fall** in temperature.

**c)** During an endothermic reaction heat energy is **released / taken in**.

**d)** During an endothermic reaction there is a **rise / fall** in temperature.

**e)** Different chemical reactions happen at the **same / different** rates.

HT

3 Draw the correct number of atoms / molecules under the equation to illustrate the change taking place.

$$2Fe + 3Cl_2 \longrightarrow 2FeCl_3$$

4 Here are some chemical reactions shown first as a word equation and then as a symbol equation. Add numbers to the symbol equations to make sure they are balanced.

**a)** Calcium + Oxygen ⟶ Calcium oxide

$$Ca + O_2 \longrightarrow CaO$$

**b)** Zinc + Hydrochloric acid ⟶ Zinc chloride + Hydrogen

$$Zn + HCl \longrightarrow ZnCl + H_2$$

**c)** Sodium hydroxide + Sulphuric acid ⟶ Sodium sulphate + Water

$$NaOH + H_2SO_4 \longrightarrow NaSO_4 + H_2O$$

**d)** Magnesium + Sulphuric acid ⟶ Magnesium sulphate + Hydrogen

$$Mg + H_2SO_4 \longrightarrow MgSO_4 + H_2$$

**e)** Calcium carbonate + Hydrochloric acid ⟶ Calcium chloride + Carbon dioxide + Water

$$CaCO_3 + HCl \longrightarrow CaCl_2 + CO_2 + H_2O$$

# Patterns in Properties

**1** The diagram shows an outline of the periodic table. Answer the questions below using only the elements shown.

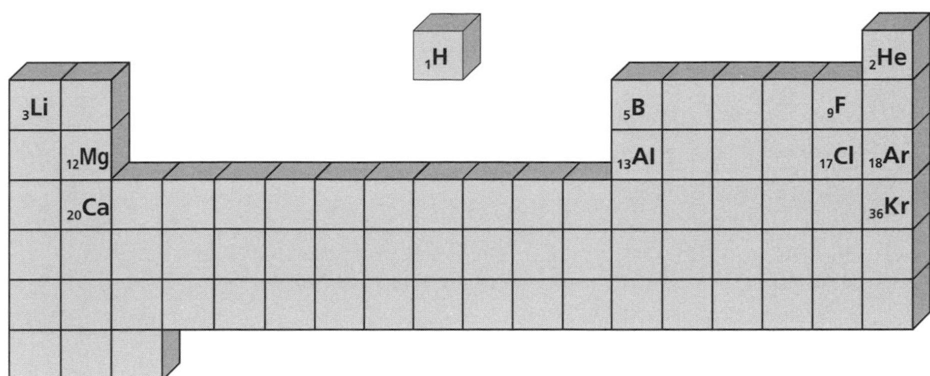


**a)** Write the name and symbol of....

**i)** a metal ................................ **ii)** a non-metal ................................

**b)** Write the name and symbol of two non-metals in the same group.

**i)** ................................ **ii)** ................................

**c)** Write the name and symbol of two metals in the same period.

**i)** ................................ **ii)** ................................

**d)** Write the name and symbol of **i)** a metal and **ii)** a non-metal in the same period.

**i)** ................................ **ii)** ................................

**e)** Write the name and symbol of **i)** a metal and **ii)** a non-metal in the same group.

**i)** ................................ **ii)** ................................

**f)** Lithium and sodium are in Group 1 of the periodic table. Explain, in terms of their atomic structure, why they have similar chemical properties.

................................................................

HT

**2** Describe two ways in which John Newlands' arrangement of the periodic table was different to the modern periodic table.

**a)** ................................................................

................................................................

**b)** ................................................................

................................................................

# Patterns in Properties

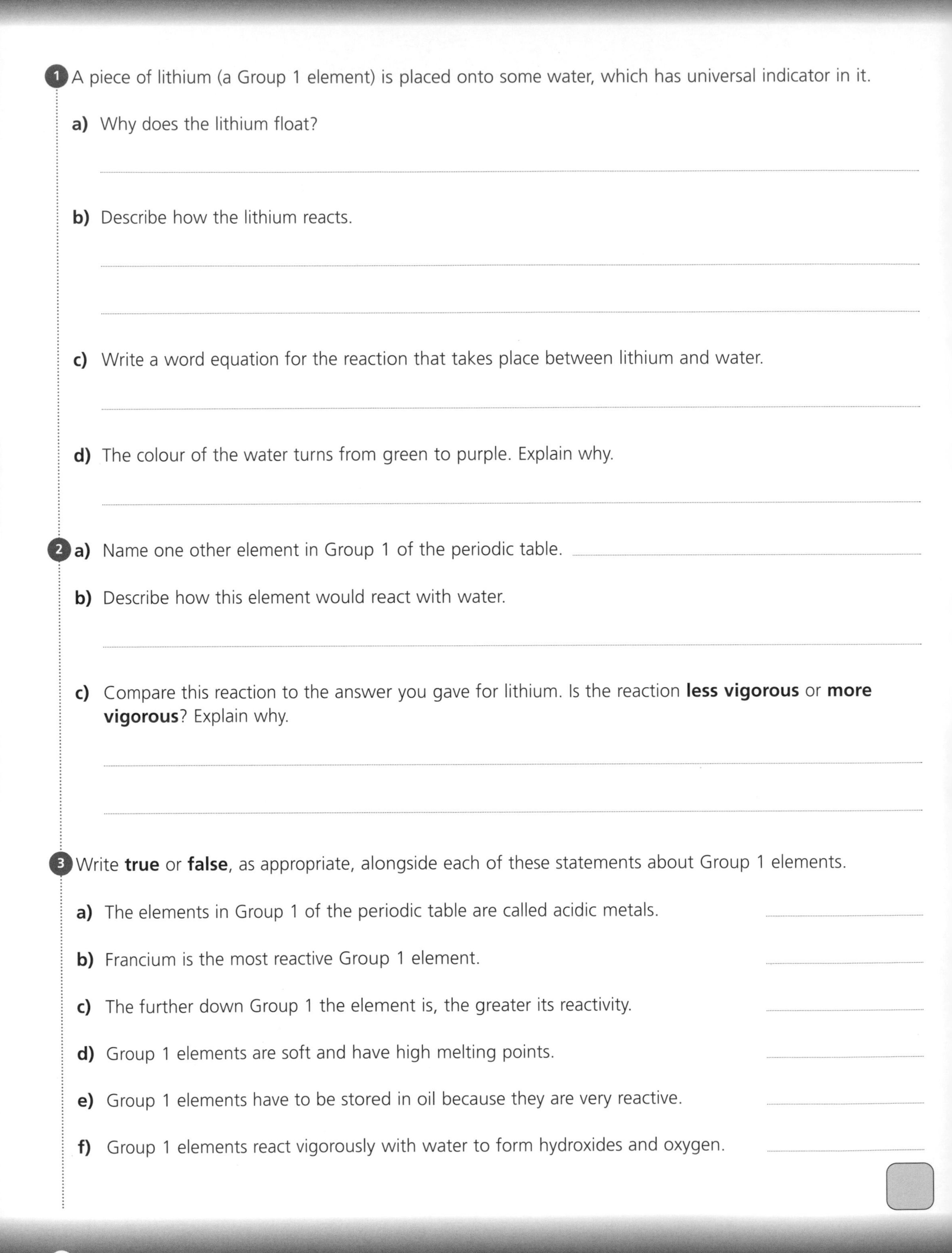

1 A piece of lithium (a Group 1 element) is placed onto some water, which has universal indicator in it.

**a)** Why does the lithium float?

**b)** Describe how the lithium reacts.

**c)** Write a word equation for the reaction that takes place between lithium and water.

**d)** The colour of the water turns from green to purple. Explain why.

2 **a)** Name one other element in Group 1 of the periodic table.

**b)** Describe how this element would react with water.

**c)** Compare this reaction to the answer you gave for lithium. Is the reaction **less vigorous** or **more vigorous**? Explain why.

3 Write **true** or **false**, as appropriate, alongside each of these statements about Group 1 elements.

**a)** The elements in Group 1 of the periodic table are called acidic metals.

**b)** Francium is the most reactive Group 1 element.

**c)** The further down Group 1 the element is, the greater its reactivity.

**d)** Group 1 elements are soft and have high melting points.

**e)** Group 1 elements have to be stored in oil because they are very reactive.

**f)** Group 1 elements react vigorously with water to form hydroxides and oxygen.

# Patterns in Properties

**1 a)** Below is an outline of the periodic table. Use a coloured pencil to shade the section of the table where the transition metals appear.

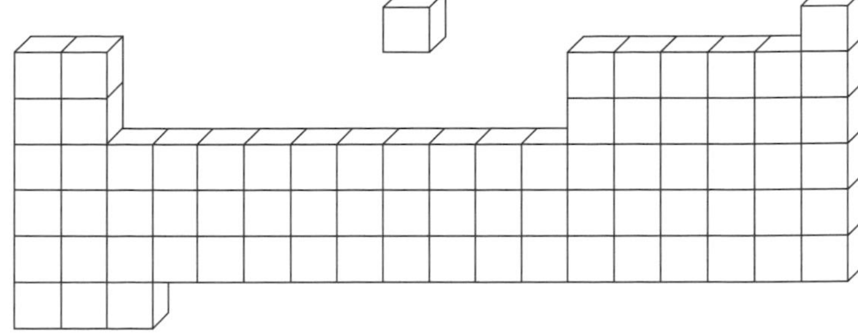

**b)** Give the names of five transition metals.

**i)** ........................ **ii)** ........................

**iii)** ........................ **iv)** ........................

**v)** ........................

**c)** List four properties that these metals have in common.

**i)** ........................ **ii)** ........................

**iii)** ........................ **iv)** ........................

**2** Transition metals have many uses. For each of the examples listed below, give one reason why the transition metal is used.

**a)** Iron for car bodies ........................

**b)** Copper for hot water pipes ........................

**c)** Copper for wiring ........................

**d)** Iron in the Haber process ........................

**3** The table below shows information about the elements known as the noble gases, Group 8 (or 0).

| Noble Gas | Atomic Number | Melting Point (°C) | Boiling Point (°C) |
|---|---|---|---|
| ........................ | 2 | -272 | -269 |
| Neon | ........................ | -248 | -246 |
| ........................ | 18 | -189 | -189 |
| Krypton | 36 | -157 | -153 |
| Xenon | 54 | ........................ | ........................ |

**a)** Complete the table, estimating the melting and boiling points of xenon.

**b)** Describe the difference in melting point and boiling point as you go down the group.

# Patterns in Properties

1 Fluorine, chlorine, bromine and iodine are all non-metals in Group 7 of the periodic table.

**a)** What is the name given to these elements? ..............................

**b)** Write down the chemical formula for…

**i)** chlorine .............................. **ii)** bromine ..............................

**c)** Describe one use of…

**i)** iodine ..............................

**ii)** chlorine ..............................

2 Tom carried out an experiment to see how halogens react with other halogen compounds. He bubbled bromine, iodine, and chlorine gas, in turn, through aqueous solutions of sodium chloride, sodium bromide and sodium iodide.

**a)** Complete the table of results alongside using a tick (✔) to indicate where a reaction took place, and a cross (✘) to indicate where there was no reaction.

| | Sodium Iodide (NaI) | Sodium Chloride (NaCl) | Sodium Bromide (NaBr) |
|---|---|---|---|
| **Bromine Gas** | | | |
| **Iodine Gas** | | | |
| **Chlorine Gas** | | | |

**b)** Why was the experiment carried out in a fume cupboard?

..............................

**c)** Write a word equation for one of the reactions shown in the table.

..............................

HT 3 Describe an experiment to investigate the reaction of chlorine with iron wool. Use a diagram to help.

..............................

..............................

..............................

..............................

..............................

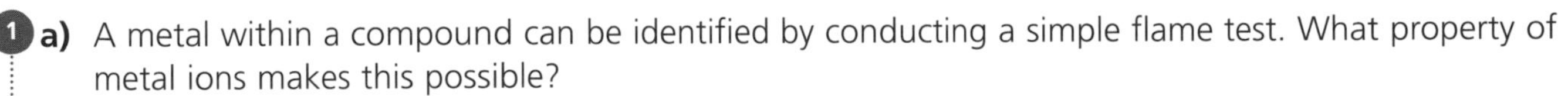

**1 a)** A metal within a compound can be identified by conducting a simple flame test. What property of metal ions makes this possible?

**b)** Describe briefly the method for carrying out a flame test.

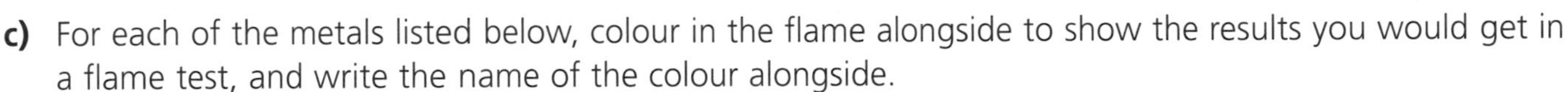

**c)** For each of the metals listed below, colour in the flame alongside to show the results you would get in a flame test, and write the name of the colour alongside.

**i)** Barium

**ii)** Calcium

**iii)** Copper

**iv)** Lithium

**v)** Potassium

**vi)** Sodium

HT **2 a)** What does it mean if the name of a metal compound ends in **-ide?**

**b)** What does it mean if the name of a metal compound ends in **-ate?**

**c)** Complete the general equations below to show the type of metal compound produced from the reactions.

**i)** Metal + Water ⟶

**ii)** Metal + Hydrochloric acid ⟶

**iii)** Metal + Sulphuric acid ⟶

# Patterns in Properties

**1** Solve the clues to complete the crossword below.

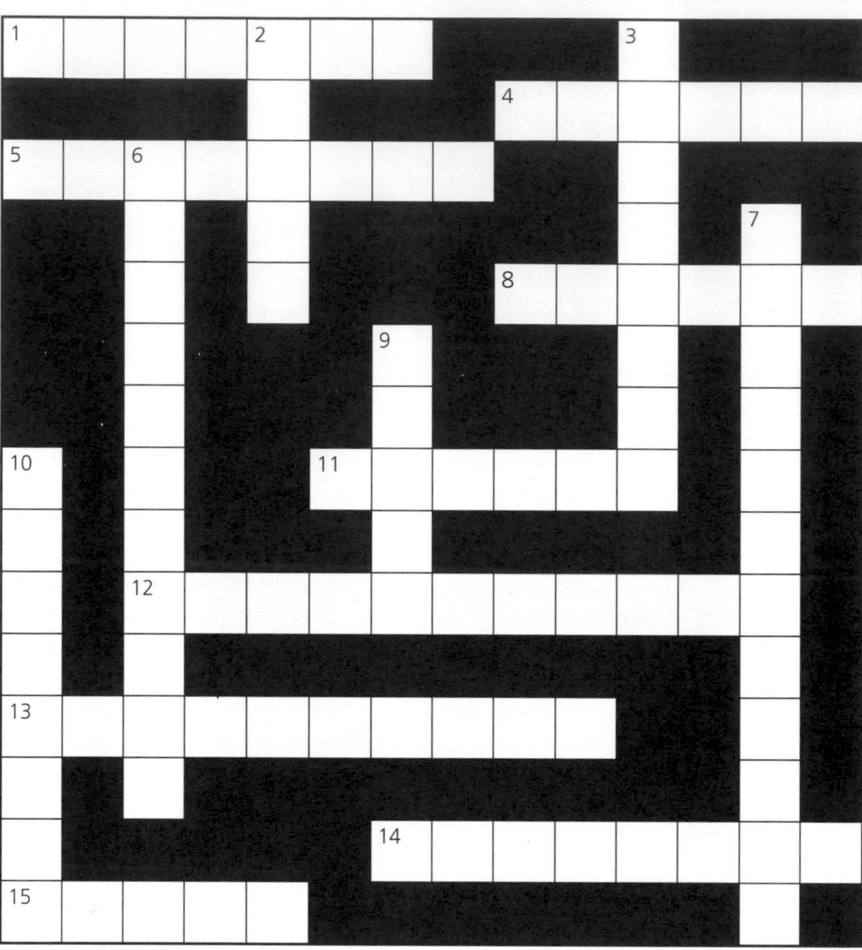


**Across**

1. The name given to the non-metals in Group 7 of the periodic table (7)
4. A positively charged particle found in the nuclei of all atoms (6)
5. The atoms of some elements cannot exist singularly, and form this type of molecule (8)
8. This number is equal to the number of protons in an atom (6)
11. A horizontal row in the periodic table (6)
12. A reaction that takes in heat energy from the surroundings (11)
13. The name given to the large block of metals between Groups 2 and 3 of the periodic table (10)
14. The uncharged particles found in almost all atoms (8)
15. The unreactive gases in Group 8 (or 0) of the periodic table (5)

**Down**

2. A vertical column in the periodic table; contains elements with similar properties (5)
3. A substance that contains two or more elements that are chemically combined (8)
6. The name given to elements in Group 1 of the periodic table (6,5)
7. This type of reaction leads to an element in a compound being replaced by a more reactive element (12)
9. Chemically unreactive (5)
10. A negatively charged particle, which orbits the nucleus of an atom (8)

# Making Changes

1 Explain what is meant by the term 'ore'.

2 A student carried out an investigation, reacting several metal oxides with carbon. The results are shown below. A tick indicates that a pure metal was obtained from the reaction.

| Metal Oxide | Magnesium Oxide | Copper Oxide | Tin Oxide | Aluminium Oxide | Lead Oxide |
|---|---|---|---|---|---|
| Reaction | ✘ | ✔ | ✔ | ✘ | ✔ |

Based on the results in the table, state the method which could be used to extract magnesium, copper, tin, aluminium and lead from their ores. Explain your answer in detail.

3 Prospectors search for gold by panning for it. Explain why gold can be obtained in this way.

4 Sodium is extracted from sodium chloride (common salt) by electrolysis. Explain why electrolysis has to be used.

5 For each of the following reactions, state whether it is an example of **oxidation** or **reduction**.

**a)** Copper oxide + Hydrogen ⟶ Copper + Water

**b)** Iron (III) oxide + Carbon monoxide ⟶ Iron + Carbon dioxide

**c)** Iron + oxygen ⟶ Iron oxide

# Making Changes

1 What is the difference between a base and an alkali?

2 What type of reaction leads to the formation of a soluble, neutral salt?

3 Complete the word equations for the following reactions.

**a)** Iron hydroxide + Sulphuric acid ⟶ ______ + ______

**b)** Copper oxide + Hydrochloric acid ⟶ ______ + ______

**c)** ______ + ______ ⟶ Calcium sulphate + Water

**d)** ______ + Hydrochloric acid ⟶ Sodium chloride + ______

**e)** ______ + ______ ⟶ Calcium chloride + Water

**f)** ______ + ______ ⟶ Sodium sulphate + ______

4 Explain, using diagrams, how you would obtain crystals of the salt formed in the reaction between copper oxide and hydrochloric acid.

5 Name two practical uses of salts.

**a)**

**b)**

# Making Changes

**1 a)** State what each of the following hazard labels means, and give one example of a substance which falls into that category.

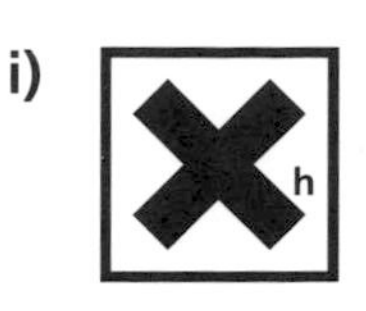

i) ......................................... .........................................

ii) ......................................... .........................................

iii) ......................................... .........................................

**b)** Draw the correct hazard symbol for each of the descriptions below:

**i)** Toxic – these substances can kill; do not swallow, inhale, or allow contact with skin.

**ii)** Irritant – these substances can cause blistering of the skin.

**iii)** Oxidising – these substances can provide oxygen, causing other substances to react or burn more fiercely.

**2 a)** What is the main distinction between natural and artificial substances?

.........................................................................................

**b)** *It is always easy to distinguish between natural and artificial substances.* Is this statement true? Explain your answer.

.........................................................................................

.........................................................................................

# Making Changes

1 What are the three main changes that food undergoes when it is cooked?

a) ............ b) ............ c) ............

2 a) In cookery, what is sodium hydrogen carbonate better known as?

............

b) Name three foods that are produced using sodium hydrogen carbonate.

i) ............ ii) ............ iii) ............

c) When it is heated, sodium hydrogen carbonate begins to break down, releasing carbon dioxide.

i) What type of reaction is this? ............

ii) Write a word equation for this reaction.

............

iii) What effect does this reaction have on foods?

............

3 Use a line to link the different types of reaction to the correct explanation.

| | |
| --- | --- |
| **a)** Hydration | Hydrogen and oxygen are removed in the form of water. |
| **b)** Thermal decomposition | Hydrogen and oxygen are added in the form of water. |
| **c)** Dehydration | The substance is broken down into simpler substances through heating. |

4 a) Draw the symbol used in equations to show that a reaction is reversible.

b) When blue copper sulphate crystals are heated, they break down into a white powder called anhydrous copper sulphate. Explain how this reaction can be reversed.

............

c) Write a word equation for this reaction.

............

# Making Changes

**1** Solve the clues to complete the crossword below, which is about different gases.

**Across**

1. The colour that damp indicator paper turns when 4 down is present (4)
7. A colourless gas that is mildly acidic (13)
9. The effect 3 down has on a glowing splint (8)
10. A colourless gas that combines with oxygen violently when ignited (8)

**Down**

2. This can be used to test for 7 across (4,5)
3. A colourless gas that aids combustion (6)
4. A colourless alkaline gas with a pungent smell (7)
5. A poisonous green gas which bleaches dyes (8)
6. 10 across produces this type of pop when ignited (7)
8. The colour that damp indicator paper turns when 5 down is present (5)

**2** Describe, with the help of a diagram, the apparatus needed to collect the product of a chemical reaction…

**a)** when the gas produced is lighter than air.

..............................................................................................................

..............................................................................................................

..............................................................................................................

**b)** when you need to measure the volume of gas produced.

..............................................................................................................

..............................................................................................................

..............................................................................................................

# Making Changes

**1** The names of ten common compounds are hidden in the wordsearch below. List their names in the space below, and for each one give an example of how it can be used.

| H | Y | D | A | N | O | A | S | S | C | A | C | N | A | U | C |
|---|---|---|---|---|---|---|---|---|---|---|---|---|---|---|---|
| Y | O | I | S | U | X | S | D | O | A | S | A | E | T | H | E |
| D | M | U | L | T | O | L | K | D | V | E | R | T | P | E | B |
| R | T | Y | O | D | N | D | O | I | I | G | B | H | H | T | C |
| O | N | P | Q | I | C | I | A | U | N | Y | O | A | O | U | A |
| C | F | W | U | N | E | C | E | M | U | S | H | N | L | I | R |
| H | T | T | A | C | I | A | H | C | P | E | Y | O | X | D | B |
| L | Q | U | M | T | I | C | C | H | K | B | D | I | I | R | O |
| O | O | O | M | P | E | I | O | L | W | N | R | C | F | O | N |
| R | X | A | O | E | J | R | U | O | M | I | A | A | P | E | D |
| I | O | V | N | E | I | T | R | R | T | H | T | C | E | L | I |
| C | A | Z | I | C | Y | I | D | I | O | G | E | I | U | O | O |
| A | H | C | A | Q | Y | C | I | D | E | I | S | D | S | B | X |
| C | E | C | L | U | F | I | E | E | V | E | H | I | M | O | I |
| I | I | Z | H | Y | Z | A | H | Y | A | M | T | I | O | N | D |
| D | S | O | D | I | U | M | H | Y | D | R | O | X | I | D | E |

**a)** ..........

**b)** ..........

**c)** ..........

**d)** ..........

**e)** ..........

**f)** ..........

**g)** ..........

**h)** ..........

**i)** ..........

**j)** ..........

# Making Changes

1 Solve the clues to complete the crossword below.

**Across**

4. Chemical compounds containing carbon, hydrogen and oxygen (13)
7. A substance formed when a base is reacted with an acid (4)
8. A reaction that chemically combines oxygen with an element or compound (9)
9. The loss of oxygen from a compound during a reaction (9)

**Down**

1. An insoluble substance that is separated from a solution during a reaction (11)
2. The word used to indicate that heat is responsible for bringing about the decomposition of a compound (7)
3. The removal of hydrogen and oxygen in the form of water (11)
4. The exothermic reaction that takes place when fuels are burnt (10)
5. To make a solution less concentrated by adding more solvent (6)
6. To add hydrogen and oxygen in the form of water (9)

# There's One Earth

1 The table below shows how the composition of the atmosphere has changed since the formation of the Earth 4.6 billion years ago.

| Composition of Atmosphere Billions of Years Ago | Composition of Atmosphere Today |
|---|---|
| Carbon dioxide 95%<br>Other gases (water vapour, hydrogen, nitrogen, carbon monoxide) 5% | Nitrogen 78%<br>Oxygen 21%<br>Carbon dioxide 0.03%<br>Other gases 1% |

**a)** Use the information in the table to describe how the atmosphere today is different from the atmosphere billions of years ago.

..........

..........

**b)** Describe the process by which the level of carbon dioxide has changed.

..........

..........

**c)** What processes have brought about the change of nitrogen in the atmosphere?

..........

..........

2 **a)** What is the name of the gas in the upper layer of the atmosphere which filters out harmful radiation from the Sun, and what element is it composed of?

..........

**b)** What type of harmful radiation does this layer shield the Earth from? ..........

3 The following statements describe some of the processes that have brought about the changes to the Earth's atmosphere. Number them **1** to **5**, to show the correct sequence of events.

**a)** Plants began to evolve and oxygen was released. ☐

**b)** New living organisms, apart from plants and bacteria, evolved. ☐

**c)** Water vapour condensed to form oceans. ☐

**d)** Ozone layer developed, filtering out harmful rays. ☐

**e)** Denitrifying bacteria act on nitrates from decaying plants, releasing nitrogen. ☐

# There's One Earth

1 **a)** Explain what is meant by the term 'global warming'.

**b)** Name three greenhouse gases.

**i)** ............ **ii)** ............ **iii)** ............

**c)** Why are greenhouse gases so called?

2 Complete the table below, listing three human activities that increase the levels of greenhouse gases in the atmosphere and are therefore likely to contribute to global warming.

| Activity | Greenhouse Gas(es) Produced |
|---|---|
| a) | |
| b) | |
| c) | |

3 Global warming would cause glaciers and the polar ice-sheets to melt. Based on this information, name three measurements that could be used to monitor global warming.

**a)**

**b)**

**c)**

4 Predictions about global warming and its effects are often based on computer models.
Suggest two reasons why these predictions might be inaccurate.

**a)**

**b)**

HT 5 In your own words, explain what the 'precautionary principle' is.

# There's One Earth

1 Name the process by which crude oil can be separated into lots of different useful substances.

.....................................................................................................

2 The diagram alongside shows a fractionating column.

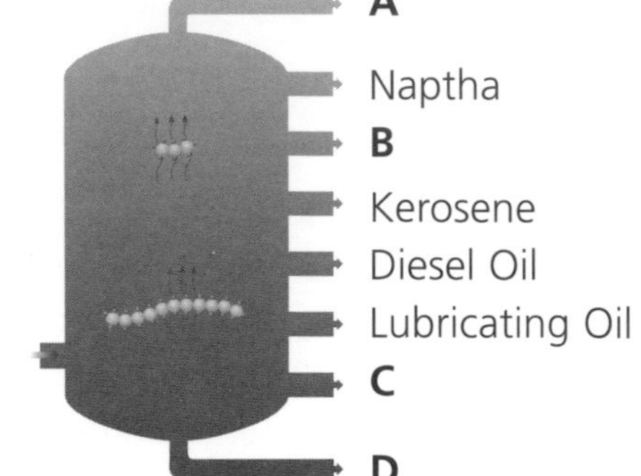


**a)** Name the fractions **A, B, C** and **D**.

**A** = .................................... **B** = ....................................

**C** = .................................... **D** = ....................................

**b)** State two physical differences between fractions **B** and **C**.

**i)** .....................................................................................................

**ii)** .....................................................................................................

**c)** What physical property is used to separate crude oil into fractions?

.....................................................................................................

**d)** Describe, in detail, the process by which the different fractions are separated in the column.

.....................................................................................................

.....................................................................................................

.....................................................................................................

3 The table alongside gives information about the boiling points of some hydrocarbons. Use this information to answer the following questions.

| Hydrocarbon | Number of Carbon Atoms | Boiling Point (°C) |
|---|---|---|
| A | 14 | 160 |
| B | 16 | 180 |
| C | 10 | 120 |
| D | 8 | 90 |
| E | 6 | 75 |
| F | 12 | 135 |

**a)** Which hydrocarbon flows most easily? ....................................

**b)** Which hydrocarbon is the least flammable? ....................................

**c)** Which hydrocarbon is the most volatile? ....................................

# There's One Earth

**1** Methane is a fuel that reacts with oxygen when burned to produce carbon dioxide and water.

**a)** What is a fuel?

**b)** Write a symbol equation for this reaction.

**c)** When methane reacts with insufficient oxygen, incomplete combustion takes place. Write a word and symbol equation for this reaction.

**d)** When methane reacts with very little oxygen, carbon is produced. Write a word equation for this reaction.

**2** Butane is a gas that is used as a fuel for camping stoves.

**a)** Write a word equation for the complete combustion of butane.

**b)** Which of the products of the combustion of butane contributes to the greenhouse effect?

**c)** The incomplete combustion of butane produces another pollutant, carbon monoxide. Why is it difficult to detect carbon monoxide?

**d)** What effect does carbon monoxide have on the human body?

**e)** Explain why it is important for rooms which contain gas fires to have a plentiful supply of fresh air.

# There's One Earth

1 The wordsearch below contains ten items that can be recycled. Put a line through each one and then list them alongside.

| | | | | | | | | | | | | |
|---|---|---|---|---|---|---|---|---|---|---|---|---|
| A | S | E | V | E | R | S | E | H | T | O | L | C |
| Y | L | S | K | I | O | L | L | E | P | D | S | A |
| C | E | U | H | O | O | T | L | B | L | O | I | R |
| K | E | N | M | O | W | E | S | Z | A | E | B | D |
| R | T | A | A | I | S | X | A | N | S | D | X | B |
| Y | S | L | G | O | M | T | P | S | T | H | O | O |
| N | E | S | A | A | R | I | A | E | I | W | E | A |
| I | R | D | Z | A | N | L | U | D | C | F | A | R |
| B | U | L | I | O | G | E | U | M | S | S | L | D |
| Y | Q | U | N | I | N | S | T | E | C | S | S | E |
| P | A | P | E | R | N | T | I | A | L | A | L | Y |
| A | M | A | S | P | N | A | R | Y | Z | Q | N | U |
| O | I | M | N | E | W | S | P | A | P | E | R | S |

a) ..................

b) ..................

c) ..................

d) ..................

e) ..................

f) ..................

g) ..................

h) ..................

i) ..................

j) ..................

2 A resource is something that is naturally occurring and of use to humans. Recycling can help to conserve resources.

**a)** Name three resources that can be conserved by recycling.

**i)** ..................

**ii)** ..................

**iii)** ..................

**b)** Give two other advantages of recycling.

**i)** ..................

**ii)** ..................

3 **a)** Suggest one reason why more companies do not recycle products.

..................

**b)** Suggest one reason why more households do not recycle waste.

..................

# There's One Earth

1 The diagram shows an arrangement for the electrolysis of a solution of sodium chloride.

- + 

a) What is the common name for sodium chloride? ……………………

b) Why must sodium chloride be in solution for electrolysis to take place?

……………………

……………………

c) Name the gas formed at...

i) the negative electrode. …………………… ii) the positive electrode. ……………………

d) Give two uses for the product formed at the negative electrode.

i) …………………… ii) ……………………

e) Give two uses for the product formed at the positive electrode.

i) …………………… ii) ……………………

f) After electrolysis has taken place, what substance remains in solution? ……………………

g) Give three uses for this substance, which remains in solution.

i) …………………… ii) …………………… iii) ……………………

2 Salt can be extracted from sea water for use in food production. The stages of this process are given below. Number them **1** to **4** to show the correct sequence of events.

a) Solar salt is dissolved in water to produce saturated brine. ☐

b) Water evaporates from the brine, leaving behind salt crystals. ☐

c) Seawater is evaporated in wide, shallow ponds to produce solar salt. ☐

d) Saturated brine is boiled in vacuum pans to purify it. ☐

3 Briefly describe how rock salt is obtained and refined to produce table salt.

……………………

……………………

……………………

# There's One Earth

1 Solve the clues to complete the crossword below.

**Across**

**3.** The means by which 11 across is started (8)
**7.** Used to describe a liquid that is thick and does not flow easily (7)
**9.** Sulphur dioxide and nitrogen dioxide mix with water in the air to form this (4,4)
**10.** This fossil fuel can be separated into many useful hydrocarbons (5,3)
**11.** The rapid oxidation reaction that takes place when fuel is burned in the presence of oxygen (10)
**12.** To collect and reuse products and materials (7)

**Down**

**1.** Poisonous (5)
**2.** The process of removing salt from water (12)
**4.** Trace substances left behind after a reaction has taken place (7)
**5.** A molecule containing only hydrogen and carbon atoms (11)
**6.** Made from biological materials, some varieties provide an alternative to petrol and diesel (7)
**8.** The type of distillation used to separate 10 across (10)

# Designer Products

1 Read the four statements carefully and place a tick beside the one that best describes smart materials.

**a)** A smart material contains microchips so that it can process data. ☐

**b)** A smart material is brightly coloured. ☐

**c)** A smart material can respond to an external stimulus. ☐

**d)** A smart material can think and has a memory. ☐

2 NOMEX® is a versatile new material which has been developed by scientists. It is very durable, flame resistant, and can withstand extremely high temperatures. It can be made into a light-weight fabric, or used to coat other materials.

Based on this information suggest two possible uses for NOMEX®.

**a)** ..............................

**b)** ..............................

3 Karen has a pair of glasses, the lenses of which get darker when she goes outside.

**a)** What is the external stimulus that causes this change? ..............................

**b)** Suggest one other possible use for this type of glass.

..............................

4 Oliver has a frying pan that has a spot in the centre, which changes colour when the pan is ready to use.

**a)** What is the external stimulus that causes this change? ..............................

**b)** Suggest one other possible use for a material that changes in this way.

..............................

5 In your own words, explain how a coat made from Gore-Tex™ stops water from getting in but is also breathable.

..............................

..............................

..............................

..............................

# Designer Products

1 a) What is nanoscience the study of?

b) One nanometre is 0.000 000 001m (one billionth of a metre). How else can this be written?

2 Why are nanoparticles added to other materials to form nanocomposites?

3 Below is a list of properties shown by different nanoparticles, and a list of some of their applications. Use a line to link each property to its appropriate application.

| Property | Application |
|---|---|
| **a)** Reflect UV radiation | Wound dressings |
| **b)** Repel water | Data storage |
| **c)** Absorb light | Anti-reflection coating for lenses |
| **d)** Kill microorganisms | Sunscreen and sun block |
| **e)** Magnetic | 'Stay-clean' coating for windows |

4 Suggest one way in which nanotechnology might be used in medicine in the future.

HT

5 Some people are concerned that nanotechnology could be misused. Describe two possible misuses of nanotechnology.

**a)**

**b)**

# Designer Products

1 a) Write **smart** or **standard** alongside each of these functions, to show what makes the difference between intelligent packaging and standard packaging.

**i)** Protects product during transit, storage and distribution ……………………

**ii)** Enhances look, taste, flavour or aroma of product ……………………

**iii)** Alerts user to changes in product or environment ……………………

**iv)** Provides product information, e.g. ingredients and weight ……………………

**v)** Keeps the components of the product together ……………………

**vi)** Communicates changing product information, e.g. product history or condition ……………………

**vii)** Prevents contamination ……………………

**b)** Based on your answers to part **a)**, explain how intelligent packaging differs from normal packaging.

……………………

……………………

2 Suggest how a smart material that responds to changes in temperature could be used in intelligent packaging.

……………………

……………………

……………………

3 Oxygen-scavenging components are added to some intelligent packaging.

**a)** What does an 'oxygen scavenger' do?

……………………

**b)** What effect does this have on the product?

……………………

……………………

# Designer Products

**1** **a)** What is the name of the microorganism that converts sugar into ethanol and carbon dioxide?

**b)** What is this process called?

**c)** Write a symbol equation for the reaction that takes place.

**2** Solve the clues about the effects of alcohol on the human body to complete the crossword below.

**Across**

1. A condition caused by a lack of 8 across (7)
5. A psychological illness which can be caused by a vitamin B deficiency (10)
8. Alcohol can decrease the levels of this important mineral (4)
9. An open sore in the stomach or intestine (5)

**Down**

2. In women, this cycle can be disrupted (9)
3. Long-term consumption of alcohol increases the risk of getting this disease, which causes tumours in organs and tissues (6)
4. Cells in this organ can be destroyed by alcohol (5)
6. Alcohol can cause .................... blood pressure (4)
7. Prolonged alcohol consumption can damage this organ, causing toxins to build up (5)

**3** As well as its physical effects, alcohol can also affect behaviour. Describe two ways in which it can do this.

**a)**

**b)**

# Designer Products

1 Below is a recipe for making Hollandaise Sauce.

*Heat 100g butter in a saucepan until it melts. In a small bowl, whisk 2 egg yolks with 1 tbsp lemon juice, a pinch of salt and a pinch of cayenne pepper. Add the melted butter to the mixture. Return mixture to saucepan and beat over very low heat until mixture thickens.*

In this recipe the egg yolk acts as an emulsifier.

**a)** What is an emulsifier?

..............................................................................................................................

**b)** Write **true** or **false** alongside each of the statements below, as appropriate, to show how the egg yolk acts as an emulsifier.

**i)** The hydrophilic part of the yolk is attracted to the lemon juice. ..............................

**ii)** The hydrophobic part of the yolk is attracted to the lemon juice. ..............................

**iii)** The hydrophobic part of the yolk is attracted to the oil. ..............................

**iv)** The hydrophilic part of the yolk is attracted to the oil. ..............................

**c)** Draw a diagram to show how an emulsifier works.

2 Explain why emulsifiers are used to make each of the following products.

**a)** Bread ..............................................................................................................................

..............................................................................................................................

**b)** Low-fat spreads ..............................................................................................................................

..............................................................................................................................

**c)** Chocolate ..............................................................................................................................

..............................................................................................................................

# Designer Products

1 Complete the crossword below.

**Across**

**2.** A colourless, flammable liquid produced by fermenting sugar (7)

**6.** Used to describe materials that can respond to external stimuli (5)

**8.** A material which combines a material, such as clay, with nanoparticles to improve its properties (13)

**Down**

**1.** A microscopic particle, with at least one dimension less than 100 nanometres (13)

**3.** A lightweight smart material that is extremely strong (6,5)

**4.** Water loving (11)

**5.** An additive that stops oil and water mixtures from separating (10)

**7.** A clear colourless alcohol of formula $C_2H_5OH$ (7)

**9.** A substance made of large molecules, formed from smaller molecules of a similar make-up (7)

# Producing and Measuring Electricity

1 What is the difference between direct current (d.c) and alternating current (a.c.)?

2 Three traces from an oscilloscope screen are shown below.

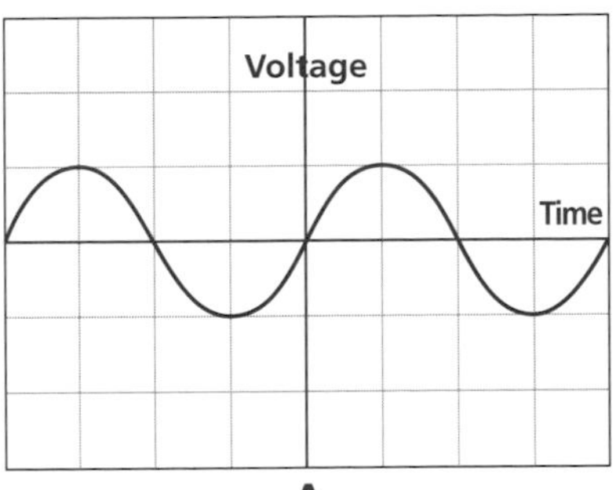


A

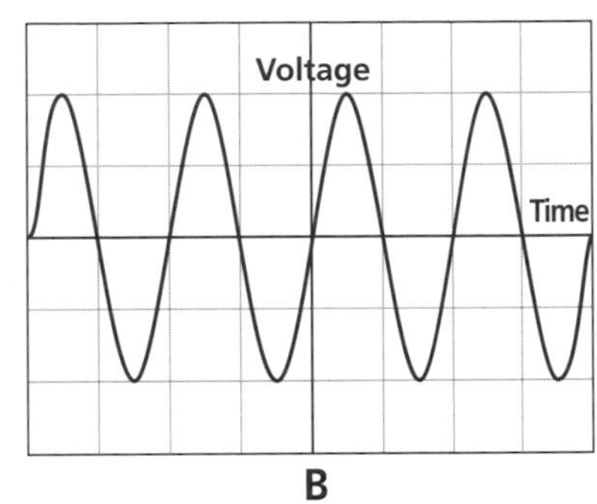


B

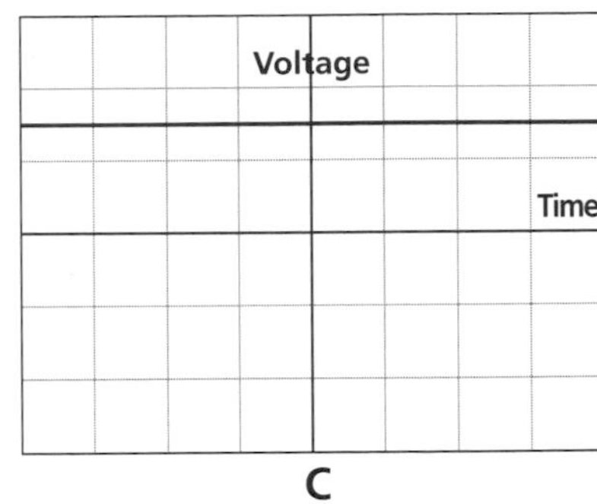


C

The settings on the oscilloscope were the same for traces A, B and C.

**a)** Which trace(s) show a direct current?

**b)** Which trace(s) show an alternating current?

**c)** If the peak voltage shown in trace A is 5V, what is the peak voltage of trace B?

**d)** If the frequency of trace B is 50Hz, what is the frequency of trace A?

**e)** On trace A, sketch an alternating current of half the frequency and three times the voltage.

3 **a)** What type of current is mains electricity?

**b)** What type of current do batteries provide?

4 Complete the table below to compare the three main types of cell (or battery).

| Type | Chemical Contents | Uses |
|---|---|---|
| Non-rechargeable dry cell | | |
| | Nickel, cadmium, lithium | |
| | | Cars, industry |

5 Are non-rechargeable batteries environmentally friendly? Give two reasons to support your answer.

# Producing and Measuring Electricity

1 The capacity of a battery is measured in amp-hours (Ah). What is the equation used for calculating capacity?

2 Use a tick ✓ or a cross ✗ to indicate whether each of the statements below is correct or incorrect.

A battery with a capacity of 25Ah will provide...

**a)** 10A of current for 2.5 hours ☐

**b)** 25A of current for 25 hours ☐

**c)** 5A of current for 5 hours ☐

**d)** 1A of current for 25 hours ☐

**e)** 50A of current for 2 hours ☐

3 A new AA battery with a capacity of 2100mAh is put into a toy train. The train runs around a track continuously and uses 0.2A. Calculate how long it will keep running before the battery goes flat.

4 Look at the simple circuit below.

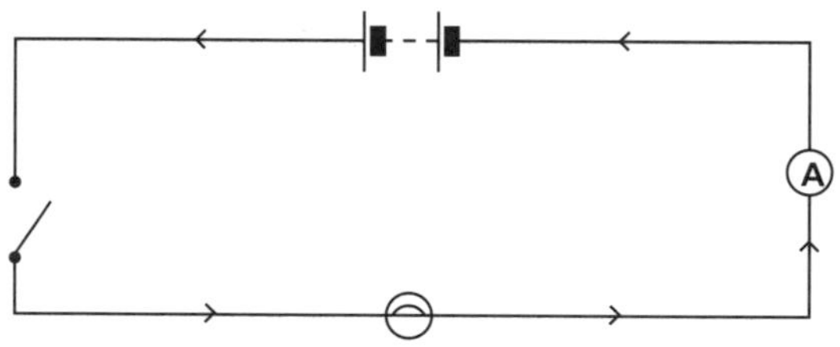

**a)** Would an electric current flow through this circuit? Explain your answer.

**b)** What is an electric current?

5 An electric current will flow through a component if there is a potential difference (voltage) across the ends of the component.

**a)** What will happen to the electric current if the potential difference across the component is increased?

**b)** What will happen to the electric current if the potential difference across the component is decreased?

# Producing and Measuring Electricity

1 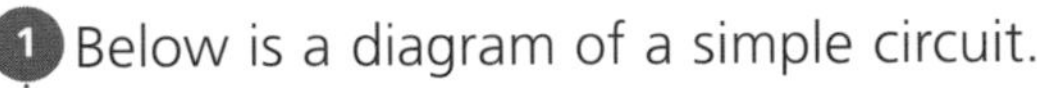
Below is a diagram of a simple circuit.

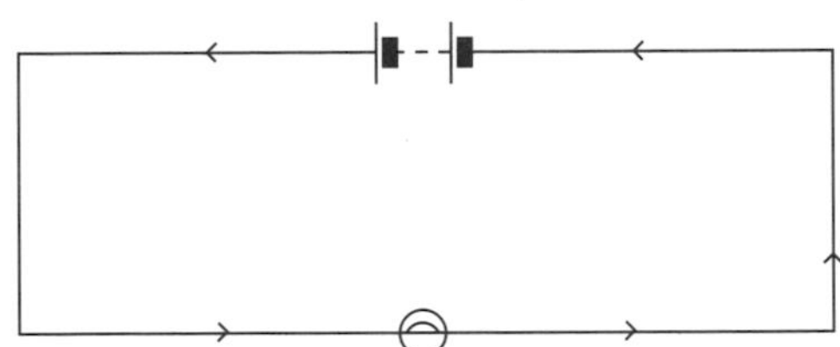

a) Using the correct symbol, draw on the diagram to show where you would connect an ammeter in this circuit.

b) Using the correct symbol, draw on the diagram to show where you would connect a voltmeter in this circuit to measure the voltage across the lamp.

c) What does an ammeter measure? ..........

d) What does a voltmeter measure? ..........

2 a) Write down the name of the component that each of the following symbols represents.

i)

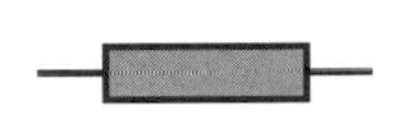

..........

ii)

..........

iii)

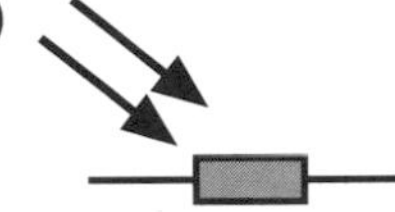

..........

b) Draw the standard symbols for each of the following components.

**i)** Battery 

**ii)** Variable resistor 

**iii)** Thermistor 

3 a) Use the box below to draw a circuit diagram, with one cell, showing two bulbs in series.

b) Use the box below to draw a circuit diagram, with one cell, showing two bulbs in parallel.

a) 

b) 

# Producing and Measuring Electricity

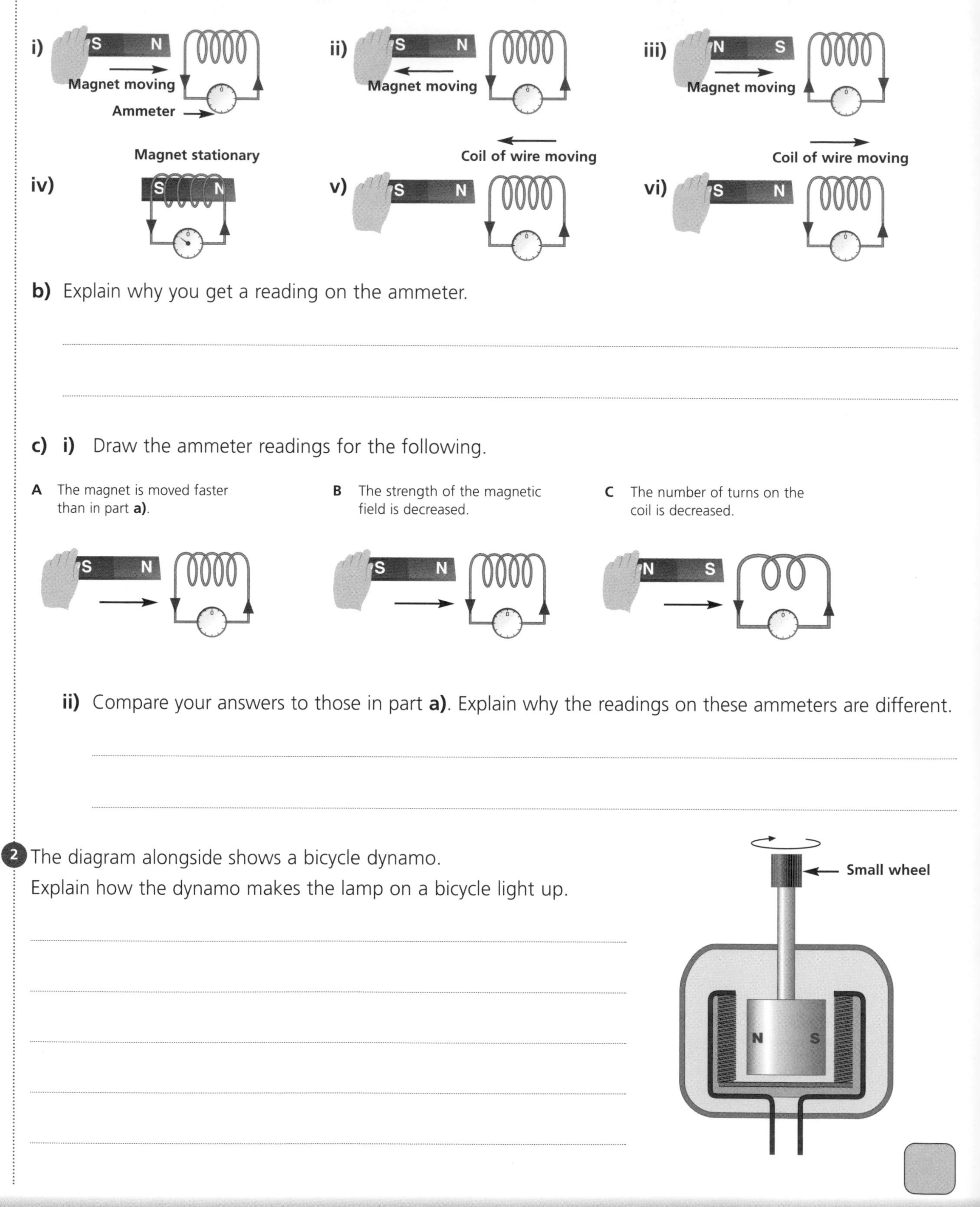

1 a) The diagrams below show a magnet and a coil of wire. The ammeter measures current. For each diagram, show the reading on the ammeter by drawing in the needle.

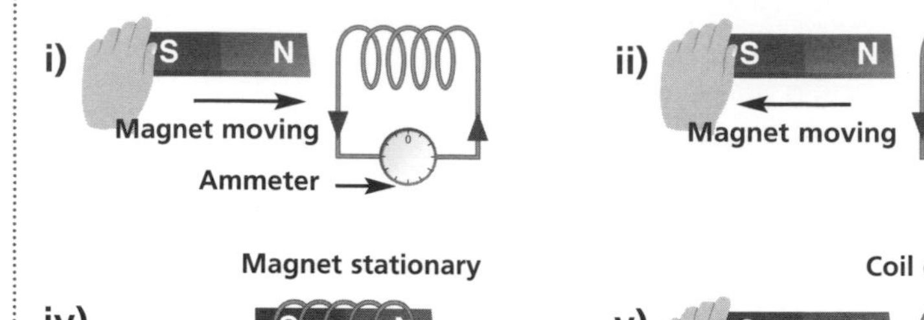


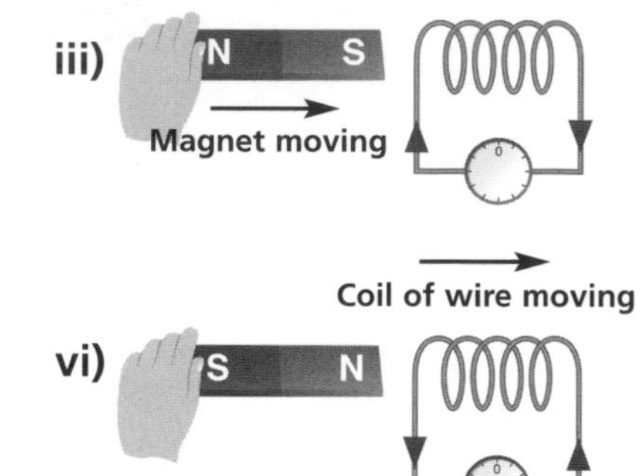


iv) Magnet stationary

v) Coil of wire moving

vi) Coil of wire moving

b) Explain why you get a reading on the ammeter.

c) i) Draw the ammeter readings for the following.

**A** The magnet is moved faster than in part **a)**.

**B** The strength of the magnetic field is decreased.

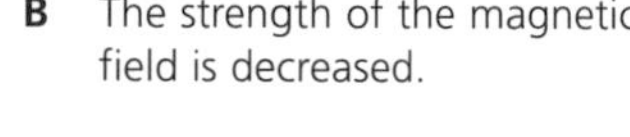

**C** The number of turns on the coil is decreased.

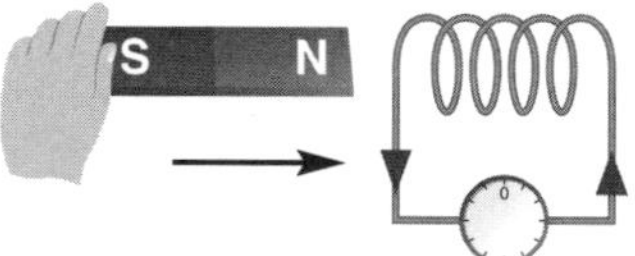


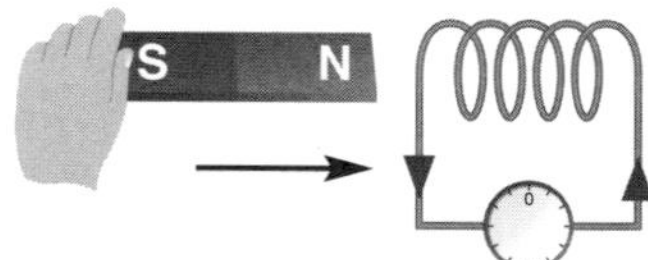


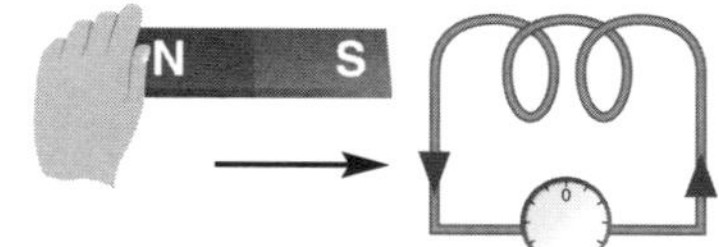


ii) Compare your answers to those in part **a)**. Explain why the readings on these ammeters are different.

2 The diagram alongside shows a bicycle dynamo.
Explain how the dynamo makes the lamp on a bicycle light up.

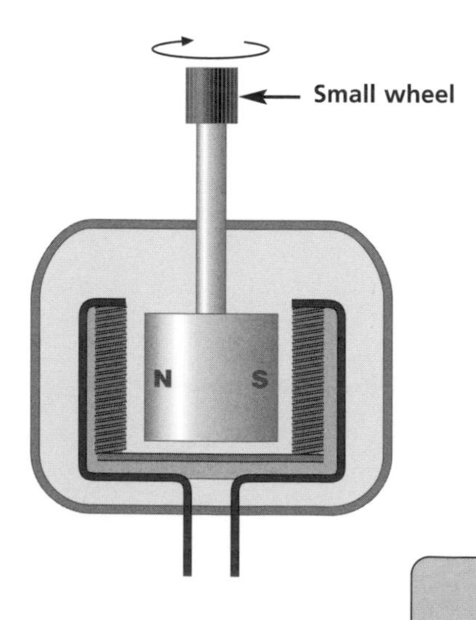

# Producing and Measuring Electricity

1 **a)** With reference to electrical components, what is resistance?

**b)** What unit is used to measure resistance? Give the full name and symbol.

2 If a light bulb is going to stop working, it usually blows the moment it is switched on. Explain, in as much detail as you can, why this happens.

3 The graph below shows how the amount of light falling on a light-dependent resistor affects its resistance.

**a)** Label the axes of the graph.

**b)** Explain the shape of the graph.

**c)** Give one example of an electrical device that uses LDRs.

4 **a)** Sketch a graph to show the relationship between resistance and temperature for a thermistor.

**b)** Give one example of an electrical device that uses thermistors.

**c)** In terms of resistance, why are thermistors unlike most other electrical components, such as light bulbs?

# Producing and Measuring Electricity

1 For the circuits shown below, each cell provides a potential difference of 1.5V. For each circuit calculate...

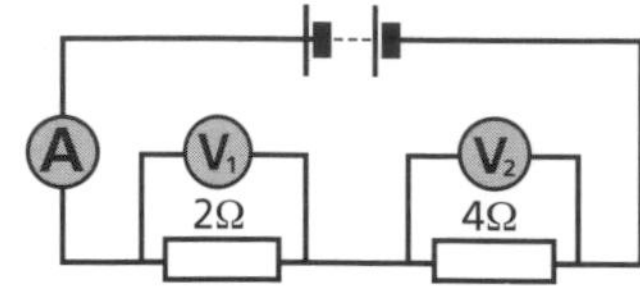


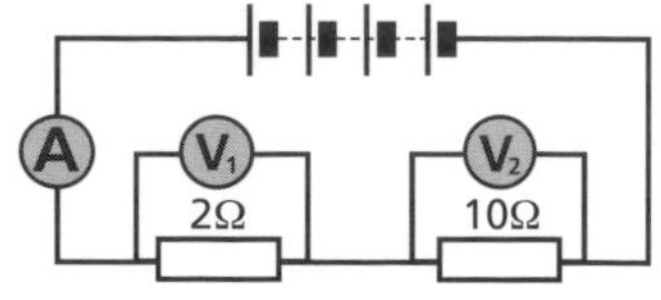


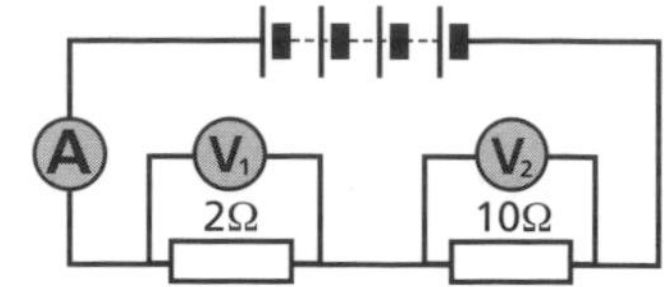


**a) i)** p.d. supplied = ..........

**ii)** Total resistance = ..........

**iii)** Ammeter reading = ..........

**iv)** $V_1$ = ..........

**v)** $V_2$ = ..........

**b) i)** p.d. supplied = ..........

**ii)** Total resistance = ..........

**iii)** Ammeter reading = ..........

**iv)** $V_1$ = ..........

**v)** $V_2$ = ..........

**c) i)** p.d. supplied = ..........

**ii)** Total resistance = ..........

**iii)** Ammeter reading = ..........

**iv)** $V_1$ = ..........

**v)** $V_2$ = ..........

2 Jill decides to investigate how the current flowing through a filament bulb changes with the potential difference across it. Jill obtained the results alongside:

| Potential Difference (V) | 0.0 | 1.0 | 2.0 | 3.0 | 4.0 | 5.0 |
|---|---|---|---|---|---|---|
| Current (A) | 0.0 | 1.1 | 1.7 | 2.1 | 2.3 | 2.5 |

**a)** Plot a current–voltage graph for this data.

**b)** Explain the shape of the graph you have drawn.

..........

..........

..........

**c)** In the space alongside, sketch a current–voltage graph to show what the shape of the graph would be if a thermistor was used instead of a bulb.

# Producing and Measuring Electricity

1 Electricity plays an important role in our everyday lives. In the space below, list everything you have in your bedroom that depends on electricity.

2 **a)** Which invention enabled the development of modern computers?

**b)** *Computers store data as a binary code*. Explain what this statement means.

**c)** Explain, in as much detail as you can, what determines the speed at which computers operate, and how they can be made to work faster.

HT

3 **a)** What is meant by the term 'superconductivity'?

**b)** What conditions are normally required for a material to achieve superconductivity?

4 **a)** What is 'Maglev' an abbreviation of?

**b)** Explain, in as much detail as you can, how a Maglev train works.

# Producing and Measuring Electricity

**1** Solve the clues to complete the crossword below.

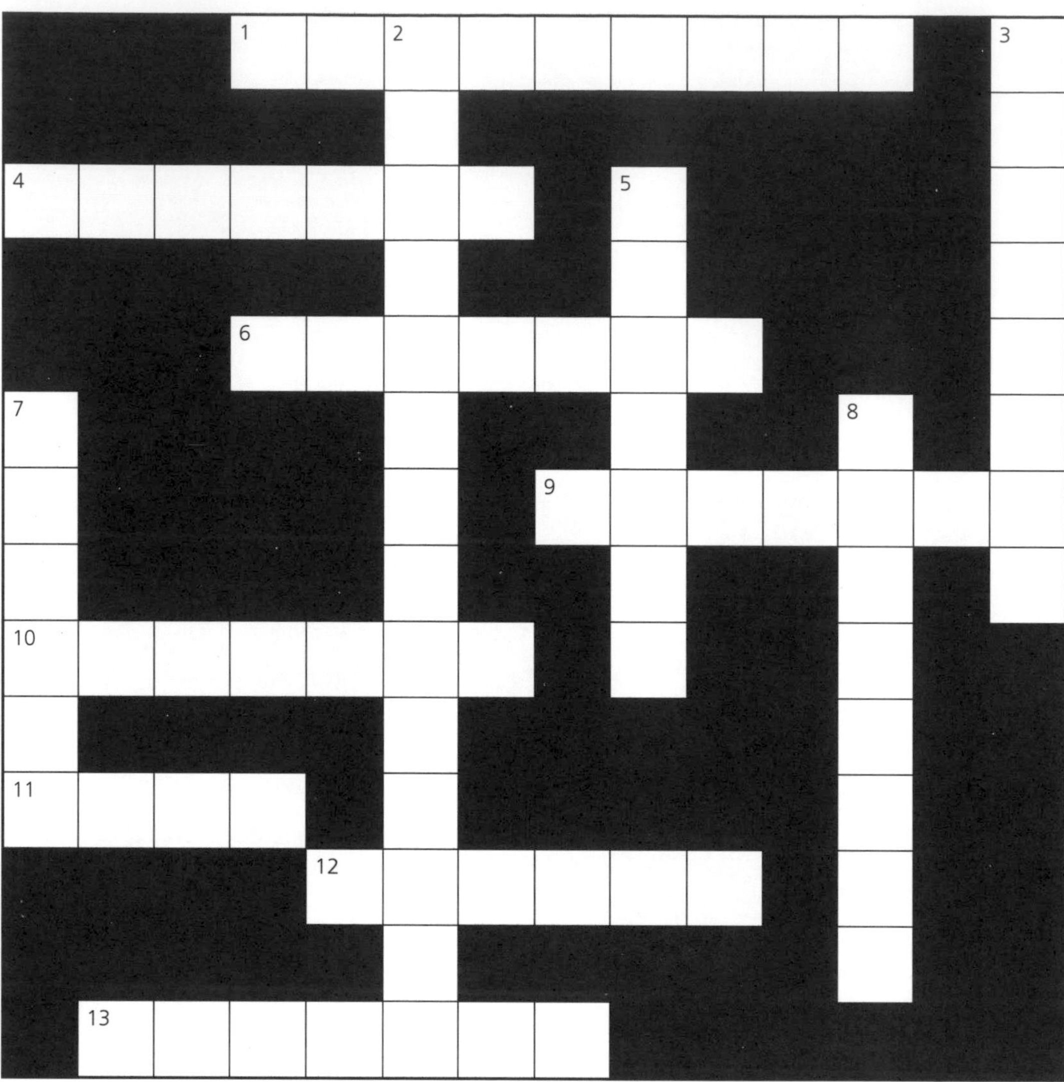


**Across**

1. A device which is able to transform light energy into electrical energy (5,4)
4. The value of the potential difference between two points (7)
6. A common device, which uses a chemical reaction to generate electricity (7)
9. A flow of electrons through a conductor (7)
10. The unit used to measure electric current (7)
11. The unit of electrical resistance (4)
12. A type of circuit in which all the components are connected in a continuous loop (6)
13. An instrument used to measure electric current (7)

**Down**

2. This type of resistor varies with light intensity (5,9)
3. A measure of how much energy 6 across can store (8)
5. A complete loop, containing electrical components, between the two terminals of a power supply (7)
7. A device that generates electricity from the rotational motion of a coil in a magnetic field or vice versa (6)
8. A component designed to produce a known resistance (8)

# You're in Charge

1 Why is electricity a secondary source of energy?

2 **a)** Arrange the following sources of energy into two groups, depending on whether they are **renewable** or **non-renewable**.

**Wind**
**Coal**
**North Sea Gas**
**Oceans (tides and waves)**
**Oil**
**The Sun**
**Geothermal Activity**
**Wood**

**Renewable** | **Non-Renewable**

**b)** What other name may be used to describe the non-renewable energy sources in part **a)**?

**c)** In your own words, explain the difference between a renewable and non-renewable energy source.

**d)** At the moment, most of the energy used by British homes and industries comes from non-renewable energy sources. Why do you think it is important for Britain to start using more renewable energy sources now?

**e)** Suggest two possible reasons why renewable energy sources are not being used more already.

**i)**

**ii)**

**f)** Name one advantage that nearly all renewable energy sources have over non-renewable sources.

# You're in Charge

3 a) Explain how a wind turbine produces electricity.

b) Wind turbines are often erected in large groups, called 'wind farms'. Why do you think this is?

c) Suggest two reasons why a rural community might object to a wind farm being built nearby.

i)

ii)

4 a) Explain how electricity is produced by a hydro-electric dam.

b) Suggest one advantage of hydro-electric energy compared to wind energy.

c) Suggest one disadvantage of hydro-electric energy compared to wind energy.

5 a) Explain how solar cells produce electricity.

b) Name two everyday products that are powered by solar energy.

i) ii)

c) Suggest one disadvantage of using solar cells to generate electricity.

# You're in Charge

HT **1** In 1818, Mary Shelley wrote a novel, called *Frankenstein*. In the book, a scientist creates a monster from different body parts, which he brings to life using electricity (from a lightning bolt).

In modern hospitals, an electrical device called a defibrillator can be used in emergency situations to try to restart a heart that has stopped beating. A pair of metal paddles are placed on the patient's chest, which deliver a jolt of direct-current electricity from the defibrillator to the heart.

**a)** How long ago did Mary Shelley write her book?

..........

**b)** Give one similarity between the procedure Mary Shelley described in her story and the use of defibrillators in modern hospitals.

..........

**c)** Give one difference between the procedure Mary Shelley described in her story and the use of defibrillators in modern hospitals.

..........

..........

**d)** Jean Jallabert and Luigi Galvani were 18th century scientists, who made important discoveries about the effects of electricity on the human body.

**i)** Explain how the work of Jallabert and Galvani might have inspired Mary Shelley to write *Frankenstein*.

..........

..........

..........

..........

**ii)** Explain how the work of Jallabert and Galvani was essential in the eventual development of modern medical treatments, like defibrillation.

..........

..........

..........

# You're in Charge

**1** The diagram shows a simple direct current motor.

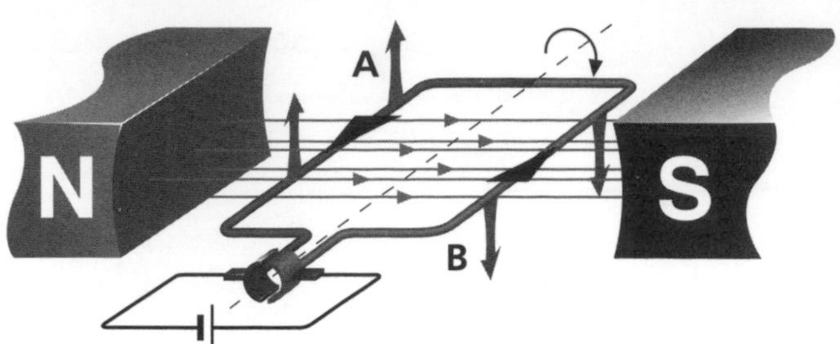


**a)** Label the following parts on this diagram.

**i)** Power supply **ii)** Carbon brushes **iii)** Commutator **iv)** Wire coil **v)** Magnets

**b)** When a current flows through the coil, side A moves up and side B moves down. Explain why.

**c)** What would happen to the rotation of the coil if the current (power supply) was reversed?

**d)** What would happen to the rotation of the coil if the magnets were reversed?

**e)** Briefly explain how the motor works. Include the purpose of the commutator in your answer.

**2** The circuit shown can be used to control the motor in a remote-controlled toy car. All cells are identical. When switch B is closed and switch A is open, the car moves forward.

**a)** What would happen to the car if switch B is open and switch A is closed? Explain your answer.

**b)** In which direction (forwards or backwards) does the car travel fastest? Explain your answer.

# You're in Charge

1 a) What determines the power of an electrical appliance?

b) What is the unit used for measuring power? Give the full written answer and correct abbreviation.

c) What two things do you need to know to calculate power?

i) ii)

d) Write the formula used for calculating power.

2 Complete the following table:

| Appliance | Electrical Power (W) | Voltage (V) | Current (A) |
|---|---|---|---|
| Iron | 920 | 230 | |
| Kettle | 2300 | | 10 |
| CD Player | 80 | 240 | |
| Vacuum Cleaner | 1400 | 230 | |
| Toaster | | 240 | 3 |

3 a) An electric motor works at a current of 3A and a voltage of 24V. What is the electrical power of the motor? (Remember to write the formula, show your working and give the correct unit of measurement.)

b) How much would it cost to run the motor for 150 minutes, if the cost of 1kWh is 7 pence?

# You're in Charge

1 Lisa likes to use her hairdryer. For every 200J of energy supplied to the hairdryer, only 80J comes out as useful energy.

**a)** What form(s) does the useful energy take?

**b)** For every 200J of energy supplied to the hairdryer, how many joules of wasted energy are there?

**c)** What form(s) does the wasted energy take?

**d)** What happens to the wasted energy?

**e)** Calculate the efficiency of the hairdryer.

2 A toy car is 40% efficient. 500J of energy is input every second. How much useful energy is output by the toy car every second?

3 Below is some data about various methods of preventing energy loss from a house. Complete the table by filling in the payback time for each method.

| Type of Insulation | Purchase Cost | Annual Savings | Payback Time (Years) |
|---|---|---|---|
| **Loft insulation** | £150 | £100 | |
| **Hot water tank jacket** | £20 | £20 | |
| **Cavity wall insulation** | £600 | £200 | |
| **Double glazing** | £2000 | £100 | |
| **Draught excluders** | £10 | £20 | |

# You're in Charge

1 a) Complete the diagram of the 3-pin plug alongside by adding any missing wires, cables and connections. Label all parts and colour-in the wires correctly.

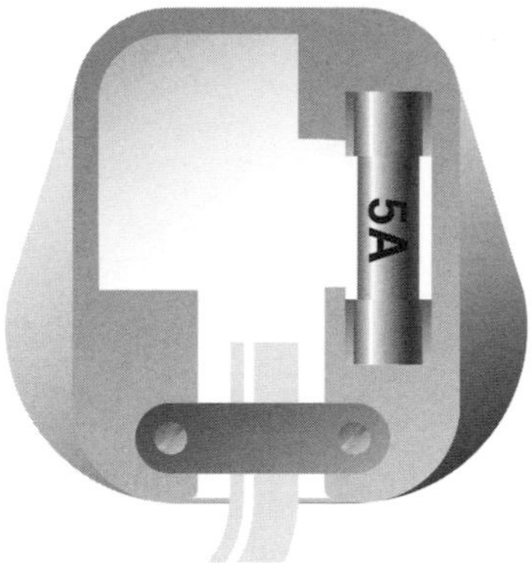

5A

b) What is the casing of the plug made of and why?

2 Look at the diagram of the 3-pin plug and write down four faults.

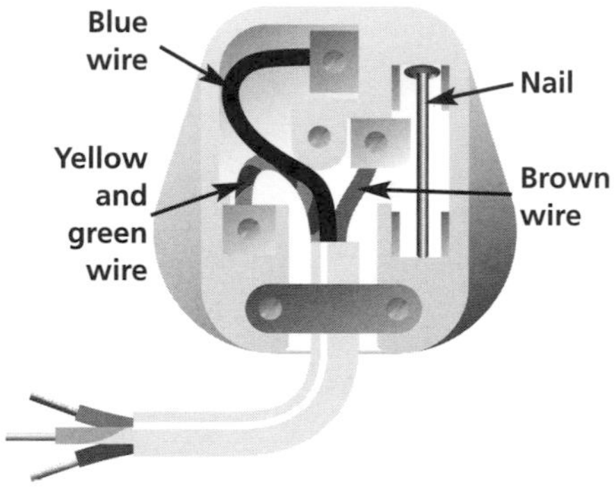

Blue wire
Nail
Yellow and green wire
Brown wire

a)

b)

c)

d)

3 a) Why does an appliance or a 3-pin plug have a fuse fitted?

b) How does a fuse work?

4 Carl buys a stainless steel toaster for his kitchen. Inside the toaster, the earth wire is connected to the outer casing. What is the purpose of the earth wire and how does it work?

# You're in Charge

1 Solve the clues and complete the crossword below.

**Across**

1. This wire is green and yellow in a standard 3-pin plug (5)
3. Used to describe an energy source that can be replaced or will not run out. (9)
4. The unit of power (5)
8. The unit of 9 across (6)
9. The ability to do work (6)

**Down**

1. The type of energy used to power most appliances in homes and industries (10)
2. A percentage measure of the amount of useful energy output by an appliance in relation to the amount of energy input (10)
5. The value of the potential difference between two points (7)
6. A device used to produce kinetic energy from a power source (5)
7. A device which is inserted into a 3-pin plug to protect an appliance from damage if the current becomes too high (4)

# Now You See It, Now You Don't

1 Jean is generating a wave by moving one end of a rope, fixed to a wall, up and down 2 times every second, as shown. In the diagram, Jean is 3m away from the wall.

3m

a) What is the frequency of the wave generated in Hertz (Hz), i.e. the number of waves per second? ..........

b) What is the wavelength of the wave generated? ..........

c) The amplitude of the wave is 0.5m. Mark the amplitude on the diagram.

d) What type of wave is this? ..........

2 The diagram alongside shows a wave travelling along a 'slinky' spring. The hand is moving backwards and forwards 3 times a second.

40cm

Hand Movement

a) What is the frequency of the wave generated in Hertz (Hz)? ..........

b) What is the wavelength of the wave generated? ..........

c) What type of wave is this? ..........

3 State two differences between transverse waves and longitudinal waves.

a) ..........

b) ..........

4 a) Complete the diagram showing what happens to white light when it hits a glass prism.

b) Visible light is part of a family of waves called the electromagnetic spectrum. The other members of the family are **X-rays**, **ultraviolet**, **radio waves**, **infrared**, **microwaves** and **gamma rays**. Arrange these waves in the correct order in the table below.

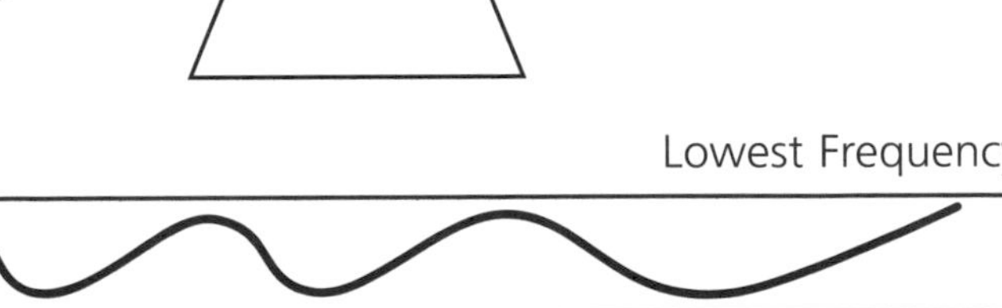


Highest Frequency — Lowest Frequency

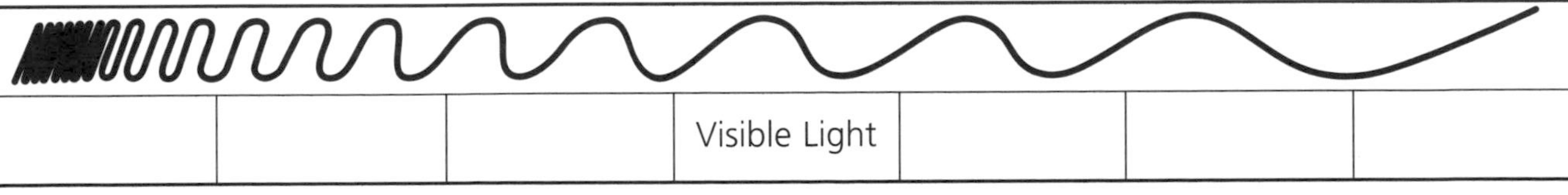

| | | | | | | |
|---|---|---|---|---|---|---|
| | | | Visible Light | | | |

# Now You See It, Now You Don't

1 a) X-rays are part of the electromagnetic spectrum. They can be very useful, but they can also be dangerous.

i) Describe one useful application of X-rays.

Pictures of bones and metals

ii) Explain why X-rays can cause damage to humans.

* some pass through the soft tissue and some are absorbed
* causes cancer

b) Microwaves are also part of the electromagnetic spectrum. Explain why microwaves and X-rays have different properties.

microwaves have a lower frequency than x-rays but larger wave length

c) Microwaves and infrared rays are both used for cooking food. Microwaves can penetrate food to a depth of 3–4cm. Infrared rays are absorbed by the surface of food.

i) Explain why microwave ovens can be used to defrost food.

Causes internal heating

ii) Explain why infrared rays are not suitable for defrosting food.

Too much exposure could burn foil as the rays are absorbed and felt as heat

2 Which type of electromagnetic radiation is used…

a) to send information to and from satellites? radiowaves

b) in sun beds (to give a sun tan)? ultraviolet

c) in remote controls for TVs and DVD players? infared

d) with fluorescent lamps to security-code electrical goods? ultrasound

3 Gamma rays can be used to kill cancer cells, however they can also be dangerous to health. Explain why.

High doses can kill normal cells and lower doses can cause cancer through the destruction or mutation of cells

# Now You See It, Now You Don't

1 a) What type of electromagnetic radiation is used to transmit signals between mobile phones and phone masts? microwaves

b) What are the dangers associated with high levels of exposure to this type of electromagnetic radiation?

can cause internal heating, body tissues that damage / kill cells. They make magnetic fields that can affect how cells work. Children are ~~much~~ more vulnerable

2 The following extract is taken from a newspaper article.

> New findings suggest that you could be frying your brain with every mobile phone call you make. Scientists have proven that mobile phones can cause a significant increase in the temperature of cells in localised parts of the brain.

a) Based on what you know about the electromagnetic radiation used by mobile phones, do you think this could be possible? Explain your answer.

Yes, as it on its microwaves. Also emission from the phone interfere with the electrical signals in the body causing headaches dizziness and cancer. They warm the brain which can slightly cause cancer

b) What the article does not tell you is that mobile phones can raise the temperature of the brain by approximately 0.1°C, but the brain's natural temperature fluctuations can be greater than this.

Taking into account this additional information, do you think that the rise in brain temperature caused by mobile phones could be dangerous? Explain your answer.

It could be. if you constantly make phone calls or talk for a long time as – to this your brain is more exposed to waves.

c) Suggest two other ways in which media reports about social-science issues might be misleading.

i) 

ii) 

3 Telephone masts give out far more radiation than mobile phones. Because of this, people often object to proposals to erect a mast close to their homes.

Draw a diagram to show why there might be a greater risk of exposure for people living in the surrounding areas than those living directly beneath the mast.

# Now You See It, Now You Don't

**1** Different types of electromagnetic radiation can be used to scan materials and produce images. Scans work by using sensors to detect when radiation is absorbed, emitted or reflected. For each of the examples below, write **absorption**, **emission** or **reflection,** as appropriate.

**a)** The use of thermal imaging equipment containing infrared sensors to track people in the dark. emmision

**b)** The use of UV light to detect forged bank notes. absorbstion

**c)** The use of iris recognition scanners to identify authorised personnel in high security situations. reflection

**d)** The use of X-rays to produce photographic images of broken bones. absorbtion

**e)** The use of microwaves to produce satellite images for monitoring rainfall. absorbtion

**2** Explain how radiographers use X-rays to produce images of a patient's bones to diagnose injuries.

The place suspected of fracture is placed in front of a photographic plate and is exposed to x-ray. The rays are absorbed by the bone but pass through the fracture and exposed the photographic plate, cleary showing where the fracture is.

**3** Police helicopters are often equipped with thermal imaging equipment, which use infrared sensors and allow the police to track people on the ground, even in the dark.

Explain how the infrared sensors work.

The detect temperature diffierence of surfaces because the high temperature, the more radiation is emmitted.

**4** **a)** What are ultrasonic waves?

they are sound waves

**b)** How are ultrasonic waves used in medicine?

they produce visual images of different parts of the body to detect problems

**c)** Give one advantage of using ultrasonic waves instead of X-rays.

there is non visible to patients, the baby, unlike x-rays

# Now You See It, Now You Don't

1 Write **true** or **false** alongside each of these statements about sound, as appropriate.

a) Sound travels in waves. true

b) Sound waves are produced when something vibrates backwards and forwards. true

c) Sound waves are transverse waves. false

d) Most humans can hear sounds in the range of 20 to 20 000Hz. true

e) Sound waves can travel through vacuums. false

f) Ultrasonic waves have frequencies less than 20Hz. false

2 a) What differences are there between analogue and digital signals?

b) List four advantages to transmitting information using digital signals.

i) no loss in quality

ii) can be handled by processors

iii) no change in the signal information during transmission

iv) second signal can be sent along the same cable at the same time

3 Digital technology has had a huge impact on the music industry. CDs and DVDs have replaced vinyl records, cassettes and video tapes.

a) Explain how data stored on CDs and DVDs are used to produce electrical signals which are then converted into sounds.

b) Suggest three other ways in which digital technology has revolutionised the music industry.

i)

ii)

iii)

# Now You See It, Now You Don't

1 An experiment was carried out where the angles of refraction and reflection were measured for eight rays of light, passing through a semi-circular glass block at different angles of incidence. The results are shown in the table:

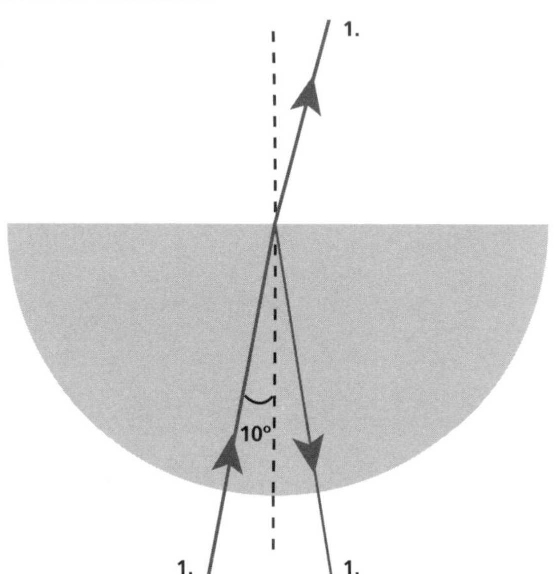


| Ray | Angle of Incidence | Angle of Refraction | Angle of Reflection |
|---|---|---|---|
| 1 | 10° | 15° | 10° |
| 2 | 20° | 31° | 19° |
| 3 | 30° | 30° | 31° |
| 4 | 40° | 75° | 40° |
| 5 | 50° | n/a | 51° |
| 6 | 60° | n/a | 60° |
| 7 | 70° | n/a | 70° |
| 8 | 80° | n/a | 80° |

**a)** Complete the ray diagram above to show the passage of all eight rays. Ray 1 has been drawn for you.

**b)** Explain why rays 5 to 8 were only reflected, and not refracted.

**c)** What factor affects the speed at which an electromagnetic wave travels through a medium?

2 The diagram shows a length of optical fibre. The critical angle for this fibre is 42°.

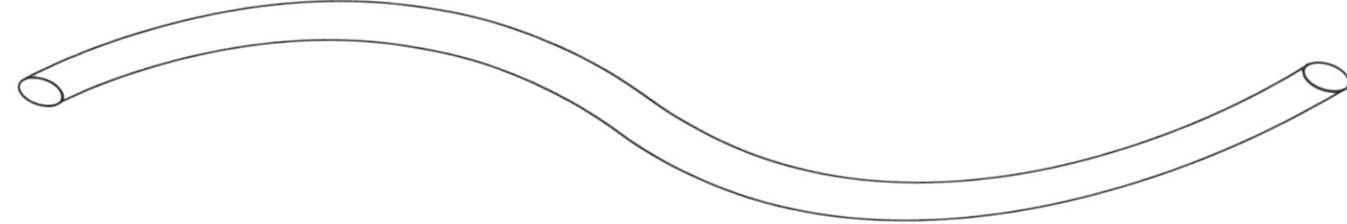

**a)** Complete the diagram by accurately drawing the passage of the ray of light through the fibre.

**b)** Optical fibres are used for telecommunications. Give two advantages of using optical fibres over conventional electrical cables.

**i)**

**ii)**

# Now You See It, Now You Don't

1 a) What two measurements do you need to know to be able to calculate wave speed?

b) What is the unit of measurement for wave speed?

c) Write the general equation used to calculate wave speed.

2 a) A sound wave has a frequency of 200Hz. Its wavelength is 1.65m. Calculate the speed of the sound wave.

b) A radio station broadcasts on a wavelength of 252m. If the speed of the radio waves is 300 000 000m/s, calculate the frequency of the radio waves.

330m/s

c) 'The Bay' radio station transmits on a frequency of 96.9MHz (96.9 million Hertz). If the speed of the radio waves is 300 000 000m/s, calculate their wavelength.

3 a) What two measurements do you need to know to calculate speed?

b) What is the unit of measurement for speed?

c) Write the general equation used to calculate speed.

3·1m

4 A fishing boat uses ultrasonic scanners to detect shoals of fish. Ultrasonic waves are sent down into the ocean from the bottom of the ship. The speed of sound in water is 1400m/s.

If it takes 0.2s for an ultrasonic wave to travel from transmitter to receiver, at what depth do the fishermen need to trawl their nets to catch the fish?

350m x 280 m

S=1400m/s d = s x t

= 1400 x 0·2

t=0·2s = 280

# Now You See It, Now You Don't

1. In your own words, describe how the structure of the Earth's surface results in the following phenomena:

   **a)** An earthquake

   **b)** A tsunami

HT

2. Why is it difficult for scientists to predict when earthquakes and tsunamis will occur?

3. The diagram shows the structure of the Earth. An earthquake occurs at point A. It produces two types of shock waves: P waves and S waves.

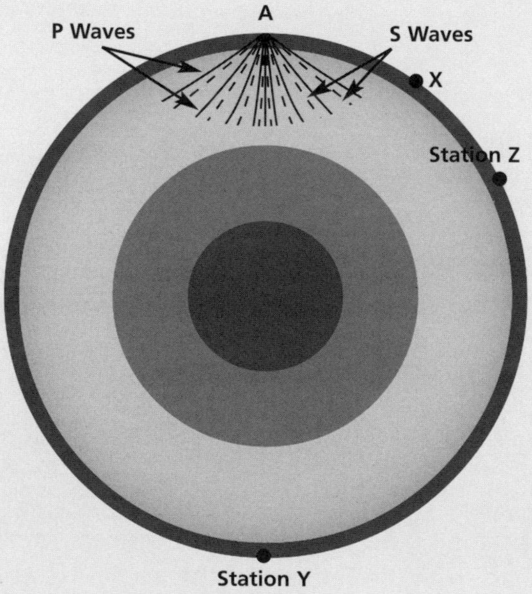

   **a)** Complete the diagram to show the path of the P waves and S waves through the Earth.

   **b)** State, with reasons, whether P waves, S waves or both will reach…

   **i)** Station Z

   **ii)** Station Y

A
P Waves
S Waves
X
Station Z
Station Y

# Now You See It, Now You Don't

1 Solve the clues to complete the crossword below.

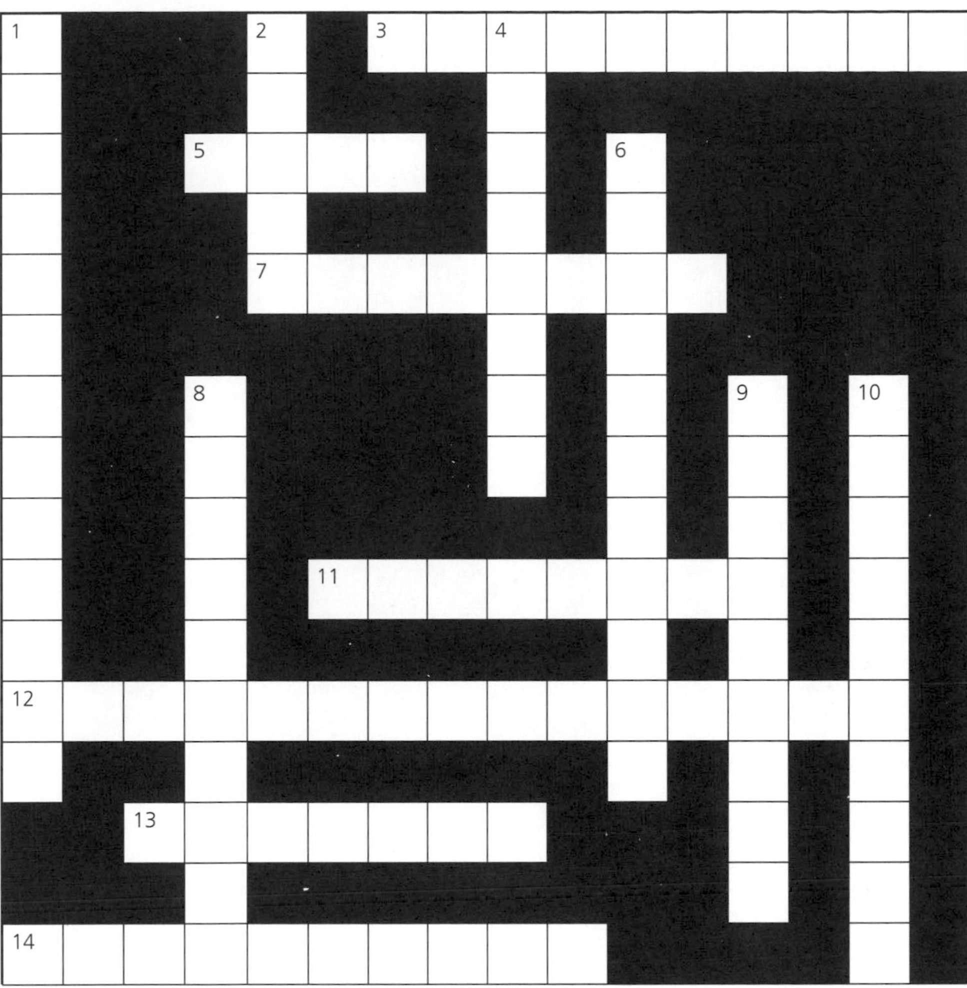


**Across**

3. Sound waves that have a frequency above 20 000Hz (10)
5. To give off radiation (4)
7. A signal that varies continuously in 9 down and / or frequency (8)
11. The angle that determines whether light is reflected or refracted at a boundary (8)
12. The type of waves that transmit energy from the Sun to Earth (15)
13. A signal that uses a binary code to represent information (7)
14. Describes a wave where the pattern of disturbance is at right angles to the direction of movement (10)

**Down**

1. Strands of glass or plastic that use totally internally reflected light to carry information (7,6)
2. These high frequency waves can be very harmful to living cells (5)
4. Describes the large plates on the Earth's surface (8)
6. Describes a substance that gives off light because it has been exposed to radiation. (11)
8. The deflection of a ray of light when it hits a boundary between two substances (10)
9. The maximum vertical disturbance caused by a wave (9)
10. This occurs when a ray of light or a wave travels from one medium into another, causing it to change in speed and direction (10)

# Space and its Mysteries

1 a) Explain why we get daytime and night-time.

..........

..........

b) On the diagram alongside, use an **X** to mark one place experiencing daytime and a **Y** to mark one place experiencing night-time.

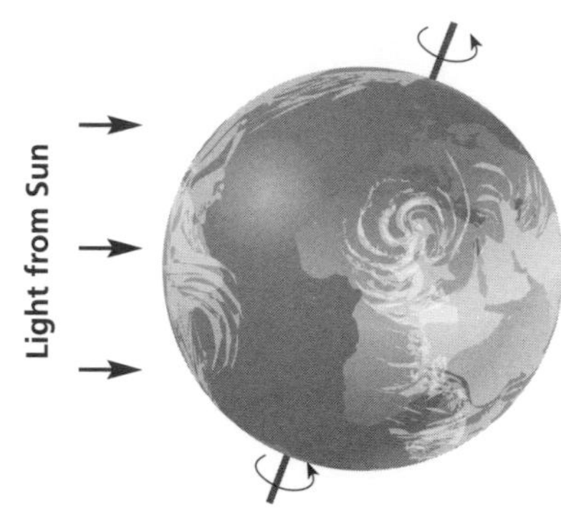


2 The diagram below shows our Solar System.

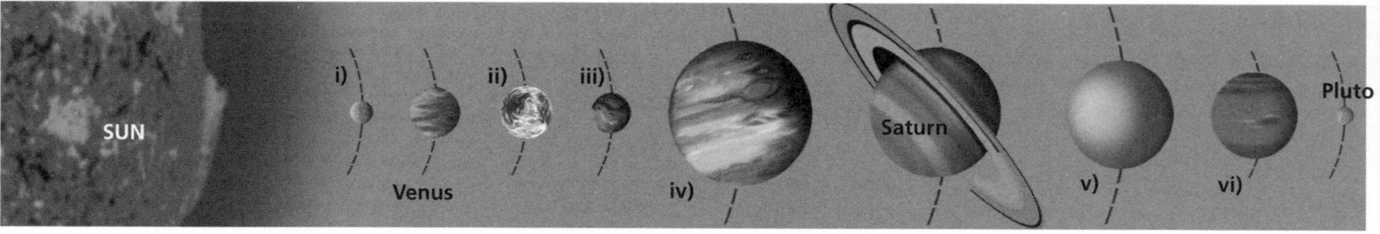


a) Six planets have not been named. Write their names below.

i) .......... ii) .......... iii) ..........

iv) .......... v) .......... vi) ..........

b) How long does it take Earth to complete one orbit around the Sun? ..........

c) Name two planets which have a shorter period of orbit around the Sun compared to Earth.

i) .......... ii) ..........

d) Why is their period of orbit less than the Earth's?

..........

e) Venus is often seen to be shining in the early evening sky. Why can we see Venus?

..........

f) Mars' period of orbit around the Sun is about 15 times less than that of Saturn. Give a reason for this.

..........

..........

# Space and its Mysteries

1 a) What is a star?

b) What is a galaxy?

c) Rewrite the following in order of size, starting with the smallest:

**Universe**, **Planet**, **Galaxy**, **Solar System**, **Star**

2 a) What is a light year?

b) What are light years used for?

3 In your own words, explain why we have seasons, and why the average temperature in summer is higher than in winter.

4 Gravity holds us firmly on the Earth's surface.

a) Although they are outside the Earth's gravitational field, why are astronauts who walk on the Moon not completely weightless?

b) In your own words, explain why astronauts orbiting the Earth experience weightlessness, even though they are still within the Earth's gravitational field.

# Space and its Mysteries

1 Explain the difference between mass and weight.

2 **a)** For each mass given, calculate its weight on Earth.

**i)** 2kg

**ii)** 10kg

**iii)** 250g

**iv)** 75g

**b)** The measurements below are the weights of four objects on Earth. Calculate each object's mass.

**i)** 250N

**ii)** 25N

**iii)** 4.5N

**iv)** 8N

**c)** An astronaut weighs 75kg on Earth.

**i)** What is his mass?

**ii)** What will his weight be on the Moon, if the Moon's gravitational field strength is 1.67 N/kg?

**iii)** What will the astronaut's acceleration of free-fall be on the Moon?

3 **a)** Write the general equation that relates force, mass and acceleration.

**b)** A motorcyclist is moving along a straight road. The total mass of the motorcyclist and motorcycle is 250kg.

**i)** The motorcyclist increases her speed from 20m/s to 30m/s in 5s. Calculate the acceleration of the motorcyclist.

**ii)** Calculate the force needed to produce this acceleration.

# Space and its Mysteries

1 In space, temperature depends on distance from the Sun.

**a)** What is the distance from the Sun to Earth? ........................................

**b)** At this distance, you would expect temperatures around 7°C. Explain why the surface of the Earth is, on average, warmer than this.

........................................

2 **a)** Why would an astronaut travelling through interplanetary space experience weightlessness?

........................................

**b)** Why would a spacecraft carrying astronauts through interplanetary space have to carry its own oxygen supply?

........................................

**c)** Describe two potential health risks to astronauts travelling through interplanetary space. For each one, suggest one way in which the risk can be reduced.

**i)** ........................................

........................................

........................................

**ii)** ........................................

........................................

........................................

3 A scientist who helps to design spacecraft thinks it would be a good idea to surround the spacecraft with water tanks. If the water isn't for drinking, why do you think he wants to do this?

........................................

........................................

HT

4 A group of scientists are designing a space station, which astronauts will be able to inhabit for long periods of time whilst conducting research in space. It is important that the space station will be able to revolve. Suggest two reasons for this.

**a)** ........................................

**b)** ........................................

# Space and its Mysteries

1 Which of the statements below best describes the law of physics relating to actions and reactions. Put a tick beside the correct one.

**a)** Every action has an equal reaction, which acts in the same direction. ☐

**b)** Every action has a greater reaction, which acts in the opposite direction. ☐

**c)** Every action has an equal and opposite reaction. ☐

2 **a)** In terms of action and reaction, explain, in as much detail as you can, how a spacecraft is propelled forward. Draw a diagram to help you explain your answer.

**b)** Why do spacecraft have to carry a 'reaction mass' into space?

HT

3 **a)** What is a black hole?

**b)** Scientists can sometimes locate a black hole in space by looking at the light from stars. Explain how this is possible.

4 To build the Hubble Telescope, scientists had to develop new technology. Some of this technology is now being used in everyday devices, like mobile phones. Describe two other products that have benefited, or been developed, from technology that originated in space research.

**a)**

**b)**

# Space and its Mysteries

1 a) Give two arguments supporting the likely existence of extraterrestrial life elsewhere in the Universe.

i) ..............................

ii) ..............................

b) Give two arguments against the likely existence of extraterrestrial life elsewhere in the Universe.

i) ..............................

ii) ..............................

2 a) Name two places in our Solar System, apart from Earth, where it may be possible to support life forms.

i) .................... ii) ....................

b) There are many ways of trying to find evidence that life does exist in other places in the Universe. Suggest a disadvantage associated with each of the following methods.

i) Sending astronauts into space to look for signs of life.

..............................

ii) Sending robots into space to bring back samples.

..............................

iii) Using robots to travel into space and send back images to Earth.

..............................

c) Describe one other method that is used to look for evidence of intelligent life in the Universe.

..............................

HT

3 List the five things that scientists look for when trying to establish whether the planets or moons in a solar system are potential homes for extraterrestrial life.

a) .................... b) ....................

c) .................... d) ....................

e) ....................

# Space and its Mysteries

1 a) Outline the Steady State Theory for the origin of the Universe.

b) Outline the Big Bang Theory for the origin of the Universe.

2 a) What type of force do bodies of mass in the Universe exert on each other?

b) As a result, explain how the total amount of mass in the Universe will determine its future.

HT

3 The diagram below shows a spectrum from the Sun and two other stars. Each of the lines represents a particular wavelength of electromagnetic radiation.

Red end of Spectrum

Sun

Star A

Star B

a) What does the spectrum of star A tell us about its movement?

b) What does the spectrum of star B tell us about its movement?

c) Explain how the evidence from spectra like those above supports the Big Bang Theory for the origin of the Universe.

# Space and its Mysteries

1 You can use the Internet, library or another secondary source to help you answer the following questions about comets.

**a)** Explain why a comet has a tail.

**b)** Why does a comet's tail point away from the Sun?

**c)** Explain why comets are not seen very often in the night sky.

**d)** Comets are only seen for a short time. Explain why this is so.

2 The two flow diagrams below show the cycle of change which occurs when a star dies. Each circle represents a different stage in the cycle.

**a)** Complete each circle in the cycle using the following words:

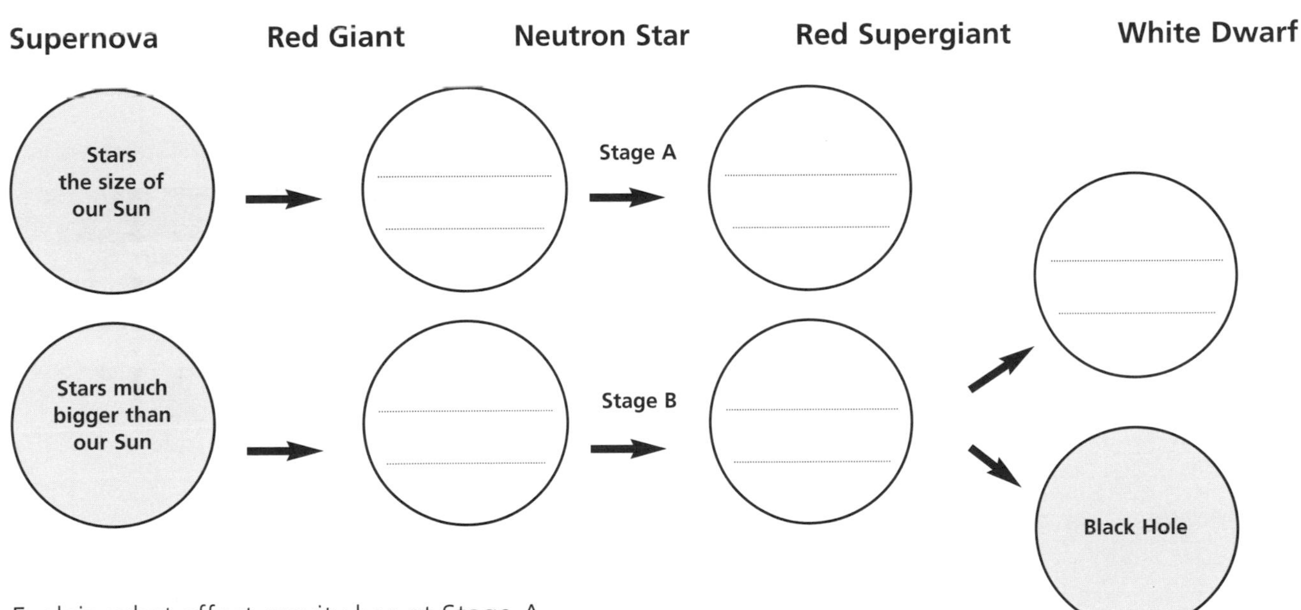


**b)** Explain what effect gravity has at Stage A.

**c)** Explain what happens during Stage B.

# Space and its Mysteries

**1** Solve the clues to complete the crossword below.

**Across**

1. The theory that the Universe began with an explosion and is expanding (3,4)
3. A group of millions of stars (6)
6. A measure of how much matter an object contains (4)
8. A layer of gases surrounding a planet (10)
12. Used to describe life outside the Earth and its atmosphere (16)
13. The name of the star at the centre of our Solar System (3)
14. The path in which a satellite moves around a larger object (5)
15. This contains everything that exists as matter and the space in which it is found (8)

**Down**

2. The pull force exerted by Earth (7)
4. Pieces of rock debris that form a band between the orbits of Mars and Jupiter (9)
5. An interplanetary body with a core of frozen gas and dust (5)
7. Relating to stars (7)
9. The clouds of gas and dust from which stars are formed (6)
10. The physical opposite of action (8)
11. Measured in newtons, this is a measure of the gravitational force acting on a mass (6)

# Notes

# Periodic Table

**Key**

| Mass number → | 1 |
|---|---|
| | **H** hydrogen |
| Atomic number (Proton number) → | 1 |

| 1 | 2 | | | | | | | | | | | 3 | 4 | 5 | 6 | 7 | 8 or 0 |
|---|---|---|---|---|---|---|---|---|---|---|---|---|---|---|---|---|---|
| | | | | | | | | | | | | | | | | | 4<br>**He**<br>helium<br>2 |
| 7<br>**Li**<br>lithium<br>3 | 9<br>**Be**<br>beryllium<br>4 | | | | | | | | | | | 11<br>**B**<br>boron<br>5 | 12<br>**C**<br>carbon<br>6 | 14<br>**N**<br>nitrogen<br>7 | 16<br>**O**<br>oxygen<br>8 | 19<br>**F**<br>fluorine<br>9 | 20<br>**Ne**<br>neon<br>10 |
| 23<br>**Na**<br>sodium<br>11 | 24<br>**Mg**<br>magnesium<br>12 | | | | | | | | | | | 27<br>**Al**<br>aluminium<br>13 | 28<br>**Si**<br>silicon<br>14 | 31<br>**P**<br>phosphorus<br>15 | 32<br>**S**<br>sulphur<br>16 | 35<br>**Cl**<br>chlorine<br>17 | 40<br>**Ar**<br>argon<br>18 |
| 39<br>**K**<br>potassium<br>19 | 40<br>**Ca**<br>calcium<br>20 | 45<br>**Sc**<br>scandium<br>21 | 48<br>**Ti**<br>titanium<br>22 | 51<br>**V**<br>vanadium<br>23 | 52<br>**Cr**<br>chromium<br>24 | 55<br>**Mn**<br>manganese<br>25 | 56<br>**Fe**<br>iron<br>26 | 59<br>**Co**<br>cobalt<br>27 | 59<br>**Ni**<br>nickel<br>28 | 63<br>**Cu**<br>copper<br>29 | 64<br>**Zn**<br>zinc<br>30 | 70<br>**Ga**<br>gallium<br>31 | 73<br>**Ge**<br>germanium<br>32 | 75<br>**As**<br>arsenic<br>33 | 79<br>**Se**<br>selenium<br>34 | 80<br>**Br**<br>bromine<br>35 | 84<br>**Kr**<br>krypton<br>36 |
| 85<br>**Rb**<br>rubidium<br>37 | 88<br>**Sr**<br>strontium<br>38 | 89<br>**Y**<br>yttrium<br>39 | 91<br>**Zr**<br>zirconium<br>40 | 93<br>**Nb**<br>niobium<br>41 | 96<br>**Mo**<br>molybdenum<br>42 | 98<br>**Tc**<br>technetium<br>43 | 101<br>**Ru**<br>ruthenium<br>44 | 103<br>**Rh**<br>rhodium<br>45 | 106<br>**Pd**<br>palladium<br>46 | 108<br>**Ag**<br>silver<br>47 | 112<br>**Cd**<br>cadmium<br>48 | 115<br>**In**<br>indium<br>49 | 119<br>**Sn**<br>tin<br>50 | 122<br>**Sb**<br>antimony<br>51 | 128<br>**Te**<br>tellurium<br>52 | 127<br>**I**<br>iodine<br>53 | 131<br>**Xe**<br>xenon<br>54 |
| 133<br>**Cs**<br>caesium<br>55 | 137<br>**Ba**<br>barium<br>56 | 139<br>**La**<br>lanthanum<br>57 | 178<br>**Hf**<br>hafnium<br>72 | 181<br>**Ta**<br>tantalum<br>73 | 184<br>**W**<br>tungsten<br>74 | 186<br>**Re**<br>rhenium<br>75 | 190<br>**Os**<br>osmium<br>76 | 192<br>**Ir**<br>iridium<br>77 | 195<br>**Pt**<br>platinum<br>78 | 197<br>**Au**<br>gold<br>79 | 201<br>**Hg**<br>mercury<br>80 | 204<br>**Tl**<br>thallium<br>81 | 207<br>**Pb**<br>lead<br>82 | 209<br>**Bi**<br>bismuth<br>83 | 210<br>**Po**<br>polonium<br>84 | 210<br>**At**<br>astatine<br>85 | 222<br>**Rn**<br>radon<br>86 |
| 223<br>**Fr**<br>francium<br>87 | 226<br>**Ra**<br>radium<br>88 | 227<br>**Ac**<br>actinium<br>89 | | | | | | | | | | | | | | | |

| | | | | | | | | | | | | | |
|---|---|---|---|---|---|---|---|---|---|---|---|---|---|
| 140<br>**Ce**<br>cerium<br>58 | 141<br>**Pr**<br>praseodymium<br>59 | 144<br>**Nd**<br>neodymium<br>60 | 147<br>**Pm**<br>promethium<br>61 | 150<br>**Sm**<br>samarium<br>62 | 152<br>**Eu**<br>europium<br>63 | 157<br>**Gd**<br>gadolinium<br>64 | 159<br>**Tb**<br>terbium<br>65 | 162<br>**Dy**<br>dysprosium<br>66 | 165<br>**Ho**<br>holmium<br>67 | 167<br>**Er**<br>erbium<br>68 | 169<br>**Tm**<br>thulium<br>69 | 173<br>**Yb**<br>ytterbium<br>70 | 175<br>**Lu**<br>lutetium<br>71 |
| 232<br>**Th**<br>thorium<br>90 | 231<br>**Pa**<br>protactinium<br>91 | 238<br>**U**<br>uranium<br>92 | 237<br>**Np**<br>neptunium<br>93 | 242<br>**Pu**<br>plutonium<br>94 | 243<br>**Am**<br>americium<br>95 | 247<br>**Cm**<br>curium<br>96 | 247<br>**Bk**<br>berkelium<br>97 | 251<br>**Cf**<br>californium<br>98 | 254<br>**Es**<br>einsteinium<br>99 | 253<br>**Fm**<br>fermium<br>100 | 256<br>**Md**<br>mendelevium<br>101 | 254<br>**No**<br>nobelium<br>102 | 257<br>**Lw**<br>lawrencium<br>103 |

→ The lines of elements going across are called periods.

↓ The columns of elements going down are called groups.